TITANIC REMEMBERED

THE UNSINKABLE SHIP AND HALIFAX

TEXT BY ALAN RUFFMAN

VISUALS FROM THE MARITIME MUSEUM OF THE ATLANTIC

FORMAC PUBLISHING COMPANY LIMITED

HALIFAX

Above: Undertaker William H. Snow at work onboard the cable ship *Minia*.
Title page visual: White Star Line promotional lithograph for *Olympic* (foreground) and *Titanic* (rear).

Formac Publishing Company Limited recognizes the support of the Province of Nova Scotia through the Department of Communities, Culture and Heritage. We are pleased to work in partnership with the province to develop and promote our culture resources for all Nova Scotians. We acknowledge the financial support of the Government of Canada through the Canada Book Fund for our publishing activities. We acknowledge the support of the Canada Council for the Arts which last year invested $157 million to bring the arts to Canadians throughout the country.

Canada NOVA SCOTIA The Canada Council for the Arts | Le Conseil des Arts du Canada

Formac Publishing Company Limited
5502 Atlantic Street
Halifax, Nova Scotia Canada
B3H 1G4
www.formac.ca

Printed and bound in Canada.

Canadian Cataloguing in Publication Data available
Ruffman, Alan.

Titanic remembered
Includes index.
ISBN 978-1-4595-0295-6

1. Titanic (Steamship). I. Maritime Museum of the Atlantic. II. Title.

First published in the United States in 2000

Contents

This symbol indicates a photo or artefact found in the Maritime Museum of the Atlantic exhibit in Halifax, Nova Scotia.

Acknowledgments

Information for this book has been accumulating for 26 years. It was Arnold and Betty Watson of the Titanic Historical Society who in 1973 first wrote on the role of Halifax, Nova Scotia, and of the Canadian vessels. They funded the 1982 visit of then-THS historians John P. Eaton and Charles A. Haas to Halifax. Their resultant "Footsteps in Halifax" special Spring 1983 issue of *The Titanic Commutator* laid out the Halifax connection, and they have repeated it in each of their subsequent books. However, little new research on the Halifax part of the story has been carried out since 1983, save for Claes-Göran Wetterholm and Peter Engberg-Klarström's identification of two victims in 1987 and the rival Titanic International Society, Inc.'s identification of four more victims in 1991.

Jack and Charlie have been constant sources over the years, as have Claus-Göran and Peter in Stockholm, Sweden and on occasion Brian Ticehurst in Southampton, England. So has Alan Hustak, of Montreal, in the past year, as first he sought my advice on aspects of his book, then as I reversed the process. Ed Kamuda, Don Lynch, and Ken Marschall of the Titanic Historical Society, Inc. have periodically helped out. Brian Cuthbertson of Halifax was very instrumental and essential in detailing Chapter 4, City in Mourning, and in providing some initial text. Howard D'Arcy, of Halifax, brought the entries in the burial and service registers of the Cathedral Church of All Saints to Brian's attention.

Marilyn Gurney at the Maritime Command Museum assisted to position Coaling Jetty No. 4 on 1912, and then on modern, maps. Anita Price and Betty Ann Aaboe-Milligan at the Dartmouth Heritage Museum sought out artefacts, as did local citizens Anita and Jane Bailey, Peter Douglass, Margaret Haley, Molly Russell, and Betty L. Thomas. Rev. Samuel Henry Prince's niece, Margaret Fulcher of Fredericton, deserves special acknowledgment for having preserved and provided his unique collection of photographs and documents. Clive Powell at the National Maritime Museum in Greenwich turned up a new manuscript, as did Laura Seymour in Calgary and Heather Wareham at the Maritime History Archive at Memorial University of Newfoundland. Steve Blasco, Bob Harmes, and Dale Buckley at the Geological Survey of Canada-Atlantic all assisted with IMAX and rusticle data while Juliette McLeod and Barbara Schmeisser at Parks Canada and Allan Clarke at the Newfoundland Museum helped with other artefacts.

Special mention is reserved for Garry Shutlak of the Public Archives of Nova Scotia. He has been a constant resource and has endured and responded to more phone calls than he will want to remember. There are also librarians and archivists at institutions around the world who all deserve praise in responding to inquiries over many years. Special recognition goes to Scott and Lisa Vollrath for preserving and sharing Scott's great-aunt Helen Tobin's summer 1920 scrapbook, made during her trip to Europe with her aunt Margaret (Molly) Tobin Brown and her sister Florence.

Credit must go to David Flemming, former Director of the Maritime Museum of the Atlantic, under whose aegis the idea of a permanent Titanic exhibit was born. It was his successor, Michael Murray and his staff, who brought it to its present success, and it was Michael who responded positively to the idea for this book. Dan Conlin, Curator of Marine History, then became the point person and has invested large amounts of time in assisting Julian Beveridge, staff of Formac Publishing, and the author with visuals, references, credits, queries, and corrections. Marven Moore and Valerie Lenethen have also been there when required. The book profited from the initial input of Scott Milsom at Formac, then Amy Black, Elizabeth Eve as copy editor, Kevin O'Reilly as designer, Julian Beveridge as photographer and Jim Lorimer as counsellor, conciliator, adjudicator and publisher. In my own office, Wendy Findley has had to wordprocess the book for over a year; hopefully she will at least be able to look at the pictures.

Introduction

Alan Ruffman has set out to relate *Titanic*'s story from a Canadian and Nova Scotian viewpoint. This is an important contribution to the ever-growing *Titanic* literature. As the final resting place of most of the recovered victims, Nova Scotia became in many eyes, the place where the liner's maiden voyage actually ended. Thanks to geography, it fell to Canadian ships to search for *Titanic* victims and Nova Scotia to receive them. Alan Ruffman's account of this chapter in the *Titanic* tragedy provides a new level of detail, putting names to the victims, their families and those who helped them. Understanding the sad events that took place in Halifax permits a deeper understanding of the aftermath of the sinking and provides a potent reminder of this region's place on the North Atlantic. The world's busiest and most dangerous sea-lane for over two centuries sweeps up the Canadian coastline for over a thousand miles, encompassing the deadly sands of Sable Island and the fog-enshrouded and sometimes iceberg-choked Grand Banks. Nova Scotia sits at the centre. The sinking of *Titanic* is one of many dramas to unfold in these waters.

The Maritime Museum of the Atlantic's Titanic *deckchair recovered after the sinking and presented by* Minia's *crew to Rev. Henry Cunningham, for his work on board the cable ship ministering to the dead.*

The Maritime Museum of the Atlantic has long displayed wooden fragments from *Titanic* which were found by mariners searching for victims. In 1996 the museum decided to feature them in an exhibit exploring the relationship between the 1912 disaster and this region, called "*Titanic*: The Unsinkable Ship and Halifax". To meet the tremendous public interest which followed the release of James Cameron's film in 1997, the museum expanded the exhibit. It has drawn record crowds and considerable acclaim. A welcome result of this attention was new exposure to the many stories at our museum which naturally flow from the *Titanic* story: from the legendary liners that called at Halifax, the city that produced the famous Cunard Line, to the thousands of other shipwrecks in Nova Scotia. The grim lessons learned in handling *Titanic*'s dead were soon needed at home when Halifax was devastated by a munitions explosion on December 6, 1917.

Alan Ruffman assisted the museum during the preparation of the *Titanic* exhibit. His career as a marine geologist provided a helpful understanding of the forces of nature and of commerce that draw people to the North Atlantic. He has measured the size of icebergs and their scour marks on the seafloor and on one occasion even tried to tow an iceberg. Alan also has a forceful interest in people and community and a relentless dedication to uncovering details. For the reader of this book, his dedication delivers a host of new facts and a clear verdict of many of the legends surrounding *Titanic* and Nova Scotia, which will help readers remember *Titanic* and her victims with a new understanding and deeper appreciation of the tragedy.

— Dan Conlin, Curator of Marine History
Maritime Museum of the Atlantic
Halifax, Nova Scotia, June 1999

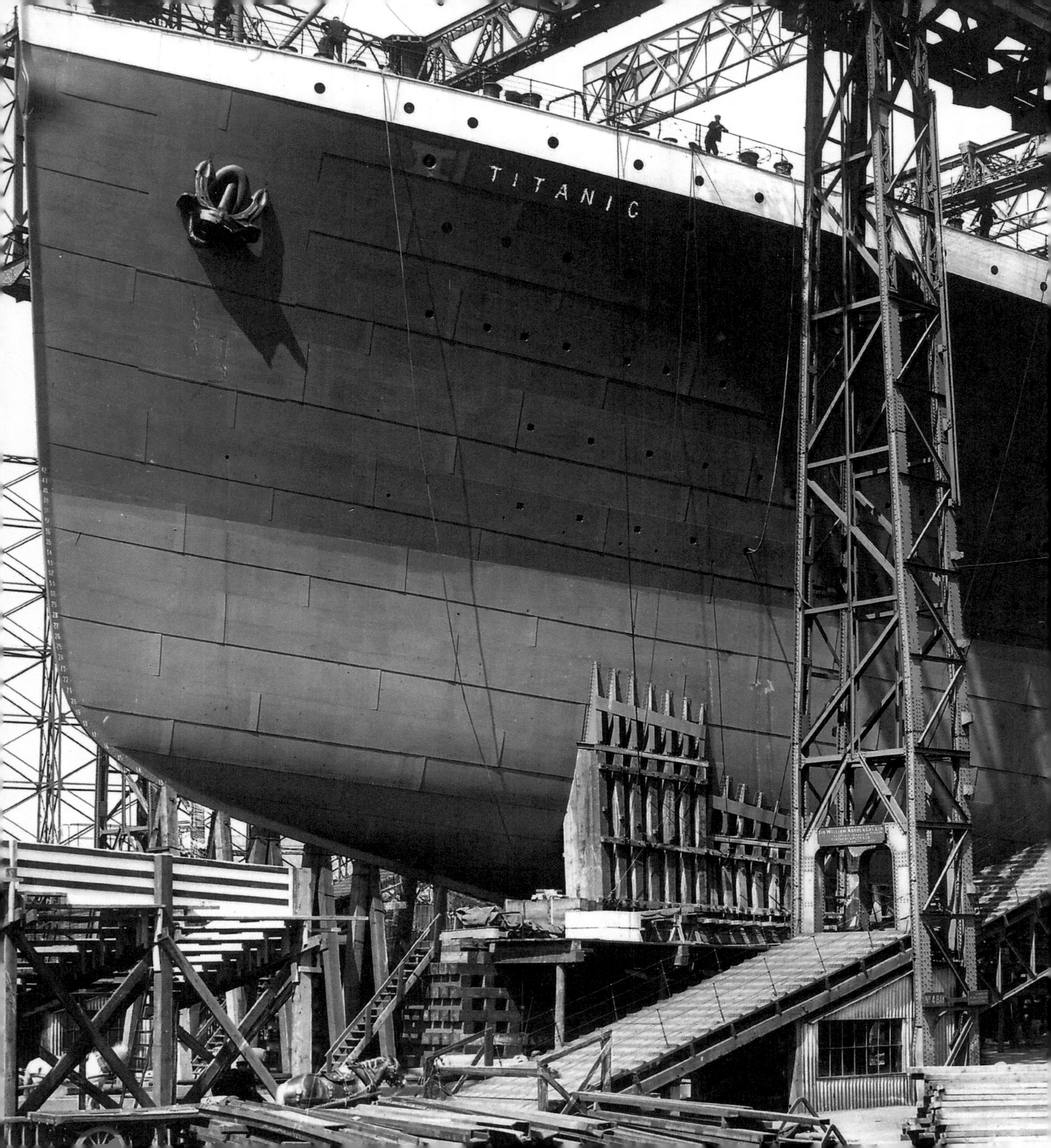
TITANIC

Chapter 1

TRIUMPH OF TECHNOLOGY TRANSATLANTIC LUXURY

Just moments before it was launched, *Titanic*, hull 401, claimed its first victim. James Dobbins was one of a large number of men working away beneath the hull in the Belfast shipyard on launch day, May 31, 1911. Twenty-two tons of tallow and soap had been laid onto the ways and a crowd of 100,000 were gathering at every possible vantage point on both sides of the Victoria Channel in the River Lagan. A swarm of workers laboured in the darkness beneath the vessel, swinging sledges to remove the forest of shores; painters came along behind them to daub the unpainted rectangle of steel left by each shore. Dobbins had been sawing out a support member when the shore collapsed, pinning the unfortunate man. His fellow workers pulled him loose and rushed him to safety from beneath the hull. It is not known if Dobbins ever regained consciousness to hear that the ship was launched smoothly on schedule just after noon. None of the 100,000 knew that Dobbins, age 43 years, was to die of his injuries the next day in the Royal Victoria Hospital.

Model of the Titanic, *completed by volunteers Gerald Wright, Jogn Green, and Bill Moore for the Maritime Museum of the Atlantic in time for the 2012 centenary of the disaster.*

The idea for the Olympic class of liners, of which *Titanic* was the newest, had originated quietly during a dinner attended by executives of Harland and Wolff Shipyards and the White Star Line. Less than a year later, in July 1908, J. Bruce Ismay, the chair and managing director of Oceanic Steam Navigation Co., owner of the White Star Line, along with numerous senior officials from the shipping line, examined the proposal for this new series of transatlantic luxury liners. Two days later the simple one-page letter of agreement was signed to initiate construction on three vessels. *Olympic* was begun in mid-December. *Titanic*'s keel was laid on March 31, 1909,

Opposite: RMS Titanic *being prepared for launch, early 1911. (RMS stood for Royal Mail Steamer)*

followed by the third sister, *Gigantic*, which was quickly renamed *Britannic* after *Titanic* was lost in April 1912.

The usual cost-plus terms were specified. Harland and Wolff built 75 vessels for the White Star Line during the 65 years they dealt with each other. All but one was built without a formal contract and all but one, on a cost-plus basis, indicating a remarkable history of handshakes and trust between the two firms.

Following the May 31 launch of the bare hull, the fitting out of *Titanic* took just over 10 frantic months. No effort was spared to improve upon the sister ship *Olympic* and to make *Titanic* the safest and most luxurious liner afloat. The improvements brought *Titanic* to 46,329 tons — the largest ship in the world. The owners gave specific instructions for a 17-inch diameter brass bell for the look-out cage on the foremast and "an iron ladder to be fitted inside the mast to give access to the cage." This was the crow's-nest from where the iceberg was first seen.

Titanic's *first-class promenade deck, portside.*

The vessel was to cater to wealthy travellers and it reeked of luxury. The servants' rooms were finished in dark mahogany with oak panelling. There were passenger elevators, for both first and second classes, a gymnasium and a small swimming pool or "plunge bath." None of these amenities were for the third-class passengers, but even their spartan quarters offered more space and privacy than usual.

Titanic and *Olympic* were the first large liners to combine a turbine engine with reciprocating engines — an innovative attempt to increase efficiency which resulted in three propellers. No expense was spared and *Titanic* was to cost over $10 million (Cdn) by the time of its first voyage — a fortune in today's dollars and much more in 1912 funds.

The date was set — Wednesday, April 10, 1912 was to be the start of *Titanic*'s first voyage to New York. In Southampton 187 first-class passengers boarded, among them J. Bruce Ismay, the Allisons, Hays and Baxters of Montreal, the Fortunes of Winnipeg, along with George Wright from Halifax, who had just made out his will the day before and left it with a London barrister. There were plenty of American millionaires and notables — Astor, Guggenheim, Stead, Straus and Widener, to name a few. The boat train from London brought second-class passengers: the Harts and their young daughter Eva, Louis M. Hoffman and young sons Michel and Edmond, and the enigmatic Baron von Drachstedt. The White Star Line's embarkation lists divided third-class passengers boarding in England into two categories. Those listed as "Other than Foreign" included the Aks, the Braunds, "going to Saskatoon, Canada," the Ford family of six, the six Goodwins, and the eleven members of the Sage family. The second category was "Scandinavian and Continental" and included the Asplunds, a Swedish family of eight bound for Worcester, Massachusetts, Vendla Heininen of Finland, Ole Olsen of Norway bound for Moose Jaw, Saskatchewan, the Swedish Alma Pålsson and four young children travelling to join their father Nils in Chicago, Jenny Lovisa Henriksson, her cousin Ellen Pettersson, and the six members of the Skoog family, all Swedes on their

Above: Titanic *at Southampton, April 1912.*
Inset: Reproduction of an early Titanic *postcard.*

way to Iron Mountain, Michigan.

Six hours later, in Cherbourg, France, 274 more passengers came on board: 142 in first class including Mrs. J.J. (Molly) Brown, 30 in second class and 102 in third class. Among the emigrants were three Greeks — Peter Lemberopoulous, Apostoles and Demetris Chronopoulos — Syrian — the Khalils and three members of the Naked family, Jamila and Elias Yarrid (possibly mis-entered as Nicola), Tamini Zabour and her daughter Hileni — and an Armenian, Nichan Krikorean, bound for Yarmouth, Nova Scotia.

Early on Thursday morning *Titanic* stopped at Queenstown (now Cobh), Ireland, for 120 more emigrants, bound mainly for New York. The passenger lists include the names Bourke, Burke, Burns, Katie Gilnagh, McCoy, Murphy and Hilda Slayter.

One first-class passenger, Francis (Frank) M. Browne ended "a trip of a lifetime" in Queenstown, a treat from his uncle, Robert Browne, Bishop of Cloyne. His day-long cruise on *Titanic* was over and he disembarked with almost 40 black-and-white photos of the voyage that now serve as a unique record of the ship's brief history. Shortly after 2 p.m. Father Browne snapped the last photo of the doomed liner as it left Queenstown Harbour, bound for New York with 2,228 passengers and crew on board. Of these, 1,523 would never arrive on the other shore.

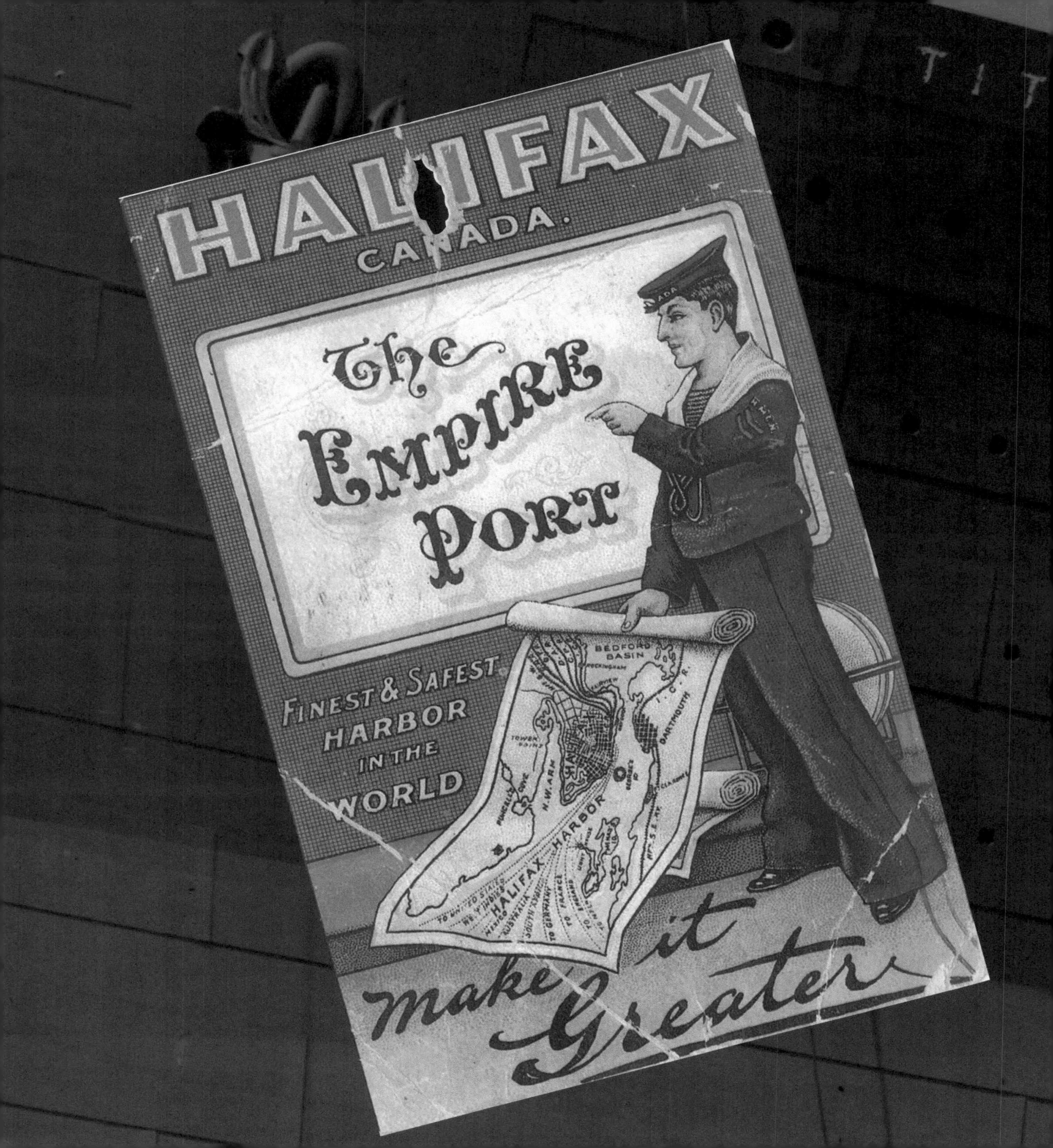
HALIFAX
CANADA.
The EMPIRE PORT
FINEST & SAFEST HARBOR IN THE WORLD
BEDFORD BASIN
DARTMOUTH
HALIFAX HARBOR
HALIFAX
Make it Greater

THE EMPIRE PORT

The transatlantic passenger trade of the late 19th and early 20th centuries was driven by the prestigious Southampton–New York traffic. The performance of each new liner on this run was the barometer against which the firms measured themselves. Prior to long-distance air travel, which developed in the late 1940s and 1950s, luxury liners were labelled "The Only Way to Cross." These ships were selling elegance and speed when compared to vessels carrying passengers on other routes.

As the three-stackers and eventually the four-stackers took over the Europe-America route, the older vessels were put in service to other parts of the British, French, German and Dutch empires. For example, the White Star Line's *Runic,* of the late 19th century, had long passed its prime and was transferred out of British ownership, ending up carrying Belgian Relief supplies during the First World War under Norwegian-ownership. This was *Imo* — one of two vessels involved in the 1917 Explosion in Halifax Harbour, Nova Scotia, one of the worst disasters in Canadian history.

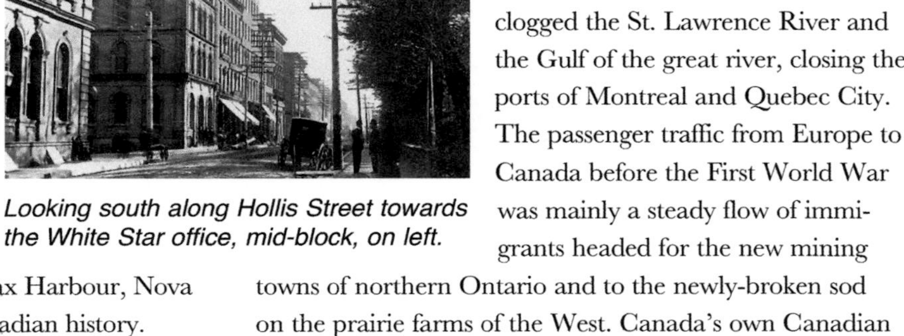

Looking south along Hollis Street towards the White Star office, mid-block, on left.

Halifax not only appointed itself as "The Empire Port," boasting the "Finest and Safest Harbour in the World," it was also home base for several of the transatlantic marine telegraph cable companies' vessels. The Commercial Cable Company operated the cable ship *Mackay-Bennett* from the Mackay-Bennett Wharf, the Western Union Company operated *Minia* from the Central Wharf, and the French cable company, La Compagnie Française des Câbles Télégraphiques, operated the cable ship *Contre-Amiral Caubet* from the Liverpool Wharf. All of these cable ship wharves were on Upper Water Street, north of Duke Street, beyond what is now called Historic Properties. The cable ships drew their crews from all over the Atlantic provinces.

Although Canada had been an independent nation since 1867, in the early 1900s it was just beginning to develop its own navy and hydrographic service. Up to this point, it had very much depended on Britain. In the first decade of the 20th century the port saw a steady stream of visiting British warships. It had become a regular stop for many of the less luxurious liners and it had served as the port of embarkation and return for Canadian volunteers in the Boer War (1899-1902).

Halifax drew even more liner traffic in the winter months when ice clogged the St. Lawrence River and the Gulf of the great river, closing the ports of Montreal and Quebec City. The passenger traffic from Europe to Canada before the First World War was mainly a steady flow of immigrants headed for the new mining towns of northern Ontario and to the newly-broken sod on the prairie farms of the West. Canada's own Canadian Pacific Steamship Company owned and operated *Empress of Britain* and *Empress of Ireland,* both two-stackers. The White Star Line maintained a presence in Halifax through the facilities of A.G. Jones and Company at 159 Hollis Street which acted as the company's agent. Halifax calls by the White Star liners *Teutonic*, *Megantic* and *Laurentic* were being advertised in local newspapers even as *Titanic*'s dead were being brought to port. In 1912 Halifax was the first landing for many of the new immigrants. Only two years later, this influx was matched by the thousands of young men leaving Canada, being sent, via Halifax, to the grim trenches of the Front in Europe.

Opposite: Period postcard , circa *1922.*

Chapter 2

At Sea

STEAMING WEST

Mr. Adolphe Saalfeld, a German-born businessman from a large chemical firm in Manchester, boarded *Titanic* in Southampton and moved into first-class cabin C106. He wrote to his wife three times on board, using Titanic stationery, and mailed the letters, addressed to "Dear Wifey," at Southampton and Queenstown.

> *I am the first man to write a letter on board ... the boat does now move & goes very steadily. It is not nice to travel alone & leave you behind, I think you will have to come next time .... I had a very good dinner and to finish had two cigars in the smoke room ... But for a slight vibration, you would not know that you are at sea .... I had quite an appetite for luncheon. Soup, fillet of plaice, a loin chop with cauliflower & fried potatoes, Apple Manhattan & Roquefort cheese, washed down with a large Spaten beer iced ... a café in the Verandah ... I shall not be able to write you again before getting to New York.*

Passengers explored the vessel, poking about until they reached the limits of their 'class.' They mapped the

FIRST CLASS PASSENGER LIST

PER

ROYAL AND U.S. MAIL

S.S. "Titanic,"

FROM SOUTHAMPTON AND CHERBOURG

TO NEW YORK

(Via QUEENSTOWN).

Wednesday, 10th April, 1912.

Captain, E. J. Smith, R.D. (Commr. R.N.R.).

W. F. N. O'Loughlin.

geon, J. E. Simpson.

Pursers { H. W. McElroy. R. L. Barker.

Chief Steward, A. Latimer.

A

	Allen, Miss Elizabeth Walton	
	Allison, Mr. H. J.	
	Allison, Mrs. H. J. and Maid	
	Allison, Miss	
	Allison, Master and Nurse	
	Anderson, Mr. Harry	
	Andrews, Miss Cornelia I.	
	Andrews, Mr. Thomas	
C101/2	Appleton, Mrs. E. D.	
B35	{ Aubert, Mrs. N. and Maid	

B

A23	Barkworth, Mr. A. H.
B58	{ Baxter, Mrs. James
B60	{ Baxter, Mr. Quigg
C6	Beattie, Mr. T.
D35	{ Beckwith, Mr. R. L. / Beckwith, Mrs. R. L.
C148	Behr, Mr. K, H.
B49	{ Bishop, Mr. D. H. / Bishop, Mrs. D. H.
T	Blackwell, Mr. Stephen Weart
A31	Blank, Mr. Henry
C7	Bonnell, Miss Caroline
C103/1	Bonnell, Miss Lily
D22/1	Borebank, Mr. J. J.

Titanic*'s first-class passenger list.* *Inset: The Maritime Museum of the Atlantic's* Titanic *replica deckchair in front of a mural of the boat deck.* *Opposite:* Titanic*'s first-class grand staircase matched this photo from* Olympic. *The newel post on the right closely resembles the artefact at the Maritime Museum of the Atlantic.*

First-class lounge on the Olympic, *a close match to the same room on the* Titanic. *The decor and carvings took their style from the palace at Versailles in France.*

corridors to the promenades or to the closest washroom. Shipboard routine had already begun to settle in as the coastline of Ireland sank below the horizon behind the fantail. Adolphe confirmed in a wireless Marconigram sent at noon Saturday that all was well from mid-ocean, "Enjoying rest. Love Adolphe." This was sent for onward transmission via SS *Minnehaha*.

Days began to melt into each other; reading, time on deck, regular mealtimes, the noon posting of the distance run (Thursday to Friday the 12th — 464 nautical miles), perhaps try out a deckchair, some conversation with new-found friends or table mates (Friday to Saturday the 13th — 519 miles), the Captain's daily tour of inspection in full dress at 10:00 a.m. with a full entourage of officers and

Titanic *carved oak panel fragment collected by* Minia's *captain. It was originally located on the archway over the forward entrance to the ship's first-class lounge. His family donated it to the Public Archives of Nova Scotia and it is now on loan to the Maritime Museum of the Atlantic.*

WHITE STAR LINE.

R.M.S. "TITANIC."

APRIL 12, 1912.

THIRD CLASS.

BREAKFAST.

OATMEAL PORRIDGE & MILK
SMOKED HERRINGS, JACKET POTATOES
TRIPE & ONIONS
FRESH BREAD & BUTTER
MARMALADE SWEDISH BREAD
TEA COFFEE

DINNER.

PEA SOUP
FRESH BREAD CABIN BISCUITS
LING FISH, EGG SAUCE
HOT POT POTATOES
STEWED APPLES & RICE

TEA.

PICKED COD
CURRY & RICE
FRESH BREAD & BUTTER
SWEDISH BREAD
JAM
TEA

SUPPER.

GRUEL CABIN BISCUITS CHEESE

Any complaint respecting the Food supplied, want of attention or incivility, should be at once reported to the Purser or Chief Steward. For purposes of identification, each Steward wears a numbered badge on the

Olympic*'s second-class smoking room.* *Inset:* Titanic*'s third-class 'Friday Fish' menu found by* Minia. *Below (left and right):* Titanic *newel post and mahogany bathroom cabinet which was recovered by an able seaman aboard the* Minia.

departmental heads. On Sunday, Divine Services were held for all classes in their respective areas. At noon whistles were tested and the officers took the noon sunshot to position the vessel — Saturday to Sunday, 546 miles. *Titanic* was now making close to 22 knots in a calm sea with clear visibility.

Meanwhile, on Sunday, Alma Cornelia Pålsson (née Berglund), age 29, was not going to forget to celebrate her son Paul's sixth birthday. Besides, her daughters, Stina Viola, now almost four, and Torborg Danira, just over eight, would remind her over and over about the importance of the occasion. There was a family celebration shared with a few of the other Swedish-speaking third-class passengers. Alma played some tunes on her mouth organ to make the event more festive.

A piece of White Star Line china, thrown overboard from an unknown liner and recovered by a fishing trawl from shallow waters on the Grand Banks of Newfoundland.

Gösta Leonard, Alma's youngest son, just over two years of age, did not fully understand the birthday event, nor indeed many of the events taking place around him. In June 1910, when he was just six months old, his father Nils had emigrated to Chicago, taking with him a studio photograph of his young family who were now finally on their way to join him in the expanding Midwest.

It was a clear moonless evening as the ship steamed west by southwest, darkness falling around it. The air temperature was much cooler. There had been a 10°F drop in temperature over the dinner hours. *Titanic* had left the warm influence of the Gulf Stream and was entering the cooler waters of the Labrador Current. It was now 33°F (0.6°C) on deck, no longer comfortable unless you were well bundled up and out of the apparent 22-knot wind created by the ship's own speed (25 mph/40.5 kph). It was easy to decide to retire early.

❸ *Detail of* Titanic *newel post carved in white oak in the "William and Mary" style.*

The ship's lamp trimmer was ordered to secure all light sources that might provide visual distractions for the crew on the bridge. A two-person watch was mounted in the crow's-nest and rotated every two hours to keep minds and eyes sharp. There was concern on the bridge that the fresh water in the outside tanks of *Titanic* would begin to freeze as the subzero seawater streamed past.

Few of the passengers or crew had ever seen pack ice or icebergs. The majority never would.

SILENT ENEMY

Icebergs are marvellous to behold in blazing sunshine. They are born at the seaward ends of glaciers that flow into the sea in Greenland and on Ellesmere, Devon and Baffin islands in northern Canada. They can tower over 100 metres into the air and reach down to scour the ocean floor over 400 metres below sea level. White with brilliant turquoise blues, occasionally with black or red-brown bands of entrained sediment from the glacier, icebergs invariably draw one's eye. If you have a camera you always find you take too many photographs of your first iceberg. This was the case with Reverend Samuel Henry Prince of St. Paul's Anglican Church in Halifax, Nova Scotia, who went out with the vessel *Montmagny* during its two searches for the bodies of *Titanic* victims. His photograph collection contains five views of the same berg as *Montmagny* slowly circled it on May 19, 1912.

At night, with no moon, and no aids such as radar, icebergs can be very difficult to see. Add in fog, snow, or a storm and the problems compound for the captain of a ship. As well, icebergs can be fickle. The main erosional process to reduce them in size occurs at the waterline where wave action cuts into them and creates a deep notch all the way around. At some point the iceberg founders as huge pieces calve off with a thunderous

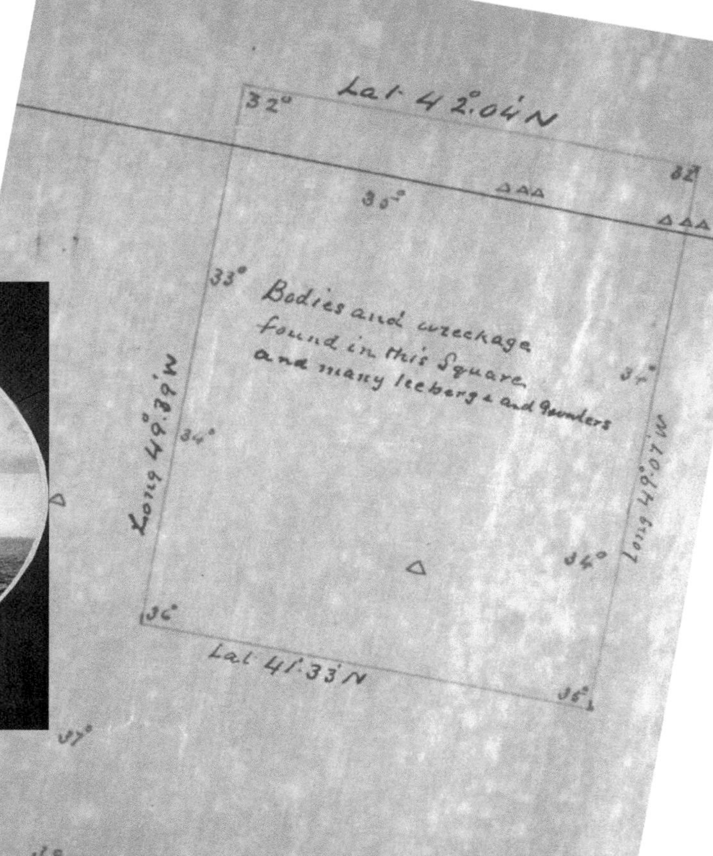

Highly eroded 'cathedral' iceberg showing several earlier water lines. Right: A portion of the Mackay-Bennett*'s original map detailing the area covered in the search for* Titanic *bodies and wreckage.*

Bow of Ivory Star *after collision with a growler, July 1974, in the Strait of Belle Isle. Left: Aerial view of an iceberg in the Spring of 1982 off Newfoundland with an underwater spur on the lower right.*

roar that sounds like cannons detonating across the sea. The sea is left littered with 'bergy bits' and growlers lying just awash. These are very dangerous to shipping until they melt away. Then the iceberg may roll to a new position to begin the process again. With each roll the berg may become more jagged and develop cathedral-like spires.

It is about a two-year journey for the icebergs to descend the ladder of the latitudes from Baffin Bay to the Grand Banks of Newfoundland. They move mainly under the influence of the cold Labrador current, passing south along the Labrador coast, and splitting to travel two routes around the Grand Banks. The inner stream stays close to the coast and the outer stream follows the eastern edge of the Banks and is directed by the shallower water east then south, past the Nose and off the Tail of the Bank.

If the spring is cold one year, and if the winter pack ice persists, then the icebergs survive longer. If there is a lot of floating first-year ice, then the waves are damped out and the icebergs do not erode as quickly and they drift much farther south than normal.

The ice of an iceberg was formed ten to twenty thousand years ago by the compression of layer after layer of snow crystals and air, making thick ice sheets that eventually began to flow outward as glaciers. When you put a piece of fresh iceberg ice in a drink, it pops and crackles as the small air bubbles release their pressure. Work on icebergs for the Hibernia oil project on the Grand Banks has shown that icebergs don't warm up inside as they float and travel south, but rather they still carry the sub-zero internal temperatures of the original ice cap. So the ice is very hard — almost rock hard — and very cold once you go only a few inches into the berg. As for climbing on an iceberg, forget it; they are as slippery as ice cubes in a drink.

If an iceberg has not rolled for a while, then the waterline erosional process along with calving may leave a nasty spur jutting out underwater. That was the case at 41°46'N on the moonless night of Sunday, April 14, 1912.

Opposite: Close-up of a massive iceberg in "iceberg alley," Northern Labrador, in July 1978.

At Sea

THE DISASTER

Captain Edward J. Smith was interviewed in the December 1908 issue of *World's Work* while still captain of *Adriatic*, four years before he took charge of *Titanic*. He was asked about the safety of the modern ocean liner. The captain exuded confidence:

> *I will not assert that she is unsinkable, but I can say confidently that, whatever the accident, this vessel would not go down before time had been given to save the life of every person on board. I will go a bit further. I will say that I cannot imagine any condition that would cause the* Adriatic *to founder. I cannot conceive of any fatal disaster happening to this ship. Modern shipbuilding has reduced that danger to a minimum.*

Captain Smith would never have to answer for those words at the official U.S. Senate and British inquiries. He could have saved his ship and its cargo of over 2,200 souls. Not at 11:40 p.m. on the night of April 14, 1912; it was too late then, but rather, 12 hours earlier when the first ice messages began to come in by wireless. He should have altered *Titanic's* course; he could have slowed down as darkness fell. The first warning had come in at 9 a.m. from *Caronia*: "West-bound steamer reports bergs, growlers, and field ice in 42°N, from 49° to 51°W, April 12." The liner *Baltic* reported by wireless at 1:42 p.m: "Greek steamer *Athenai* reports passing icebergs and large quantities of field ice today in latitude 41°51'N, longitude 49°52'W."

Opposite: Lifejacket of James McGrady, body 330, recovered by the Newfoundland sealing vessel Algerine.

In fact the ether was full of ice messages that day. *Amerika* reported passing two large icebergs in 41°27'N, 50°08'W, on April 14. *Californian* sent a message to the captain of *Antillian* which *Titanic* also recorded: "Six-thirty p.m., apparent ship's time; latitude 42°03'N, longitude 49°09'W. Three large bergs 5 miles to southward us. Regards Lord." From *Mesaba* to *Titanic* and all eastbound ships: "Ice report in latitude 42°N to 42°23'N, longitude 49° to 50°30'W. Saw much heavy pack ice and great number of icebergs. Also field ice. Weather good, clear."

This was received at 9:40 p.m, just two hours before disaster was to strike. Unfortunately it appears the last ice warning was put aside by the Marconi wireless operators who were very busy handling passengers' messages to Cape Race. The warning never got to the bridge.

Regardless, Captain Smith had lots of time and many reasons to adjust to a more southerly course to avoid the ice. But he did not. He trusted his lookouts, as was the practice of the day.

Wireless key similar to that used at Cape Race. Right: Pocket watch of Swedish Titanic *victim Mauritz Ådahl, stopped at 2.35 a.m. shortly after the sinking.*

In the words of Second Officer Charles H. Lightoller at the British inquiry, under Lord Mersey, "It was calm, perfectly calm" on the evening of Sunday, April 14, 1912. This worked to the disadvantage of the two lookouts, Frederick Fleet and Reginald Lee, who came on watch in the crow's-nest at 10:00 p.m. There was no swell and there were no waves breaking and splashing at the base of the iceberg to make it more visible through the darkness. There was also no breeze to move the overly chilly microclimate that exists around pack ice and icebergs. This microclimate can be a warning to a lookout who, feeling the chill, senses ice in the immediate vicinity, even if it cannot be seen.

The impending disaster was signalled by three blows to the gong in the crow's-nest and an immediate telephone message to the bridge. Fleet shouted, "Iceberg right ahead." Instinctively, First Officer William Murdoch ordered the wheel "Hard-a-starboard" which caused the ship to alter its course to port, or to the left. Engines were reversed but at 22 knots there was not enough time for *Titanic* to respond. The protruding underwater spur of the iceberg raked down the starboard side of the hull. The damage was done.

With the first five watertight compartments opened to the sea, the ship was doomed. It was just a matter of time before it sank. Two hours and forty minutes. One and a half minutes for each of the 110 hours of the fatal maiden voyage. Those 160 minutes have been played out and replayed many times since 1912, with the instant books of the day, Walter Lord's famous book, *A Night to Remember,* and the film that followed, Broadway plays, songs at camp — "Oh It Was Sad When the Great Ship Went Down" — an IMAX film documentary, a Discovery Channel television film, and most recently James Cameron's epic *Titanic* feature film that, at over three hours, outlasts the sinking itself by half an hour.

The horror and the heroism of the sinking have been well recorded. By the time word of the disaster filtered down to third class and Alma got the four children dressed and struggled them all up the now-tilting stairs to the upper decks, it was too late. The lifeboats were gone. She could only gather her brood of children in a quiet space and resort to the harmonica in her pocket to calm them as the angle grew more frightening and death became more inevitable.

A few people suspected that the ship had broken in two just before the stern sank, from the survivors' reports. However, it was 73 years before this was known conclusively. What was known at 2:20 a.m. Monday, April 15, 1912 was that:

...the angle had increased
From eight on to ninety when the rows
Of deck and porthole lights went out, flashed back
A brilliant second and again went black.
Another bulkhead crashed, then ...
the liner took
Her thousand fathoms journey to her grave.

.....

And out there in the starlight, with no trace
Upon it of its deed but the last wave
From the Titanic *fretting at its base,*
Silent, composed, ringed by its icy broods,
The gray shape with the paleolithic face
Was still the master of the longitudes.

from "The Titanic," E.J. Pratt (1935)

Above: Robert Hunston, wireless operator. Inset: A cut-up souvenir piece of J.J. Astor's lifejacket recovered by Mackay-Bennett. *Opposite: Hunston's log of wireless messages received at Cape Race on the night of* Titanic*'s sinking and donated to the Maritime Museum of the Atlantic by his daughter, Molly Russell.*

The! of

Titanic Disaster as viewed from Cape Race by Wireless. April 14th 1912.

EST

10.25 pm E.S.T. J. C. K. Godwin on watch hears Titanic calling CQD giving position 41.44 N 50.24 W. about 380 miles SSE of Cape Race.

.35 Titanic gives corrected position as 41.46 N 50.14 W. A matter of 5 or six miles difference. He says "have struck iceberg".

.40 Titanic calls Carpathia and says "we require immediate assistance".

.43 Titanic Gray on duty

.45 gives same information to Californian

— Caronia giving Titanic's position circulates same information broadcast to Baltic and all ships

.55 who can hear him. CH on duty.

11.00 Titanic tells German steamer "Have struck iceberg and sinking".

.25 Titanic continues calling for assistance and giving position.

Establish communication with Virginian here and give him all information re Titanic telling him inform Captain immediately. OK.

Olympic asks Titanic which way latter steering. Titanic replies "We are putting women off in boats".

Virginian informs us he is 200 miles from scene of disaster.

Virginian says he is now go[ing to] assistance. Titanic

...ic meanwhile

...ating ...

...ays last he heard of

... at 12.27 when

...als were blurred an...

...uptly.

... boats working among

...ative to Titanic

...thing more heard from

...om New York asking

This is followed by

...ore chiefly from

... many ships askin...

...ht news commences to

... ships stating

... picked up 20 boats

... word of any more bei...

FALSE HOPES

Even as *Carpathia* was speeding to New York with *Titanic* survivors, the owners of the White Star Line were in denial over the loss. Not only did Phillip A.S. Franklin, Vice President of the International Mercantile Marine Company, American owner of the White Star Line, deny to the press on Monday, April 15, that *Titanic* had sunk, or even could sink, but three times that day, at about 11:15 a.m., 1:10 p.m. and 4:30 p.m., he participated in setting up train transportation for the passengers from Halifax, where it was announced they would be landing. Benjamin Campbell, who was in charge of traffic for the New York, New Haven and Hartford Railroad Co., testified before the U.S. Senate Committee of Commerce investigating the disaster on May 16 and placed in the record Franklin's telegram sent to Campbell, written after their last afternoon telephone call.

Lifejacket of Titanic *survivor Madeleine Astor.*

> *Confirming our conversation over the telephone to-day, this is to advise you that we shall be glad if you will bill us for the transportation of all the Titanic's passengers to whom you give passage from Halifax to New York or any intermediate point and for all the meals of the passengers en route.*
>
> *We understand from our conversation with you that you were providing 30 sleeping cars and 3 dining cars for the first and second class passengers, numbering approximately 610, and a sufficient number of day coaches for 710 third-class passengers, and a sufficient number of baggage cars for all classes.*

The rigid class distinctions persisted even in the White Star Line's rescue plans. By 7:00 p.m. the truth was realised and the railroad equipment, some of which had begun to move northward from Boston, was recalled. It is not surprising that the Boston papers got it wrong.

The Boston Daily Globe for Monday evening, April 15, 1912 headlined:

TITANIC GOING
TO HALIFAX
Damaged Liner's
Passengers Taken Off
WILL LAND AT HALIFAX
NY, NH & HRR Officials Notified to
Prepare Five or More Trains

Of course the stories were cruelly wrong. By 7:30 p.m. on Monday, the Cunarder *Carpathia* had long since made its full-steam early morning run of three-and-a-half hours to arrive at the scattering of 19 upright lifeboats floating in the lonely sea. Barely 700 survivors were huddled against the cold, to be systematically taken into *Carpathia*'s warmth over the next four-and-a-half hours; three corpses that had been placed in the Collapsible A during the night were recovered by the liner *Oceanic* on May 13 and buried at sea. Four more died of exposure by the time the lifeboats reached *Carpathia* and they were buried from her decks at 4:00 p.m. on Monday afternoon, and another crewmember, seaman W.H. Lyons, was to die on board *Carpathia* and would be buried the next morning at 4:00 a.m., before any of the passengers were up and about.

Captain Arthur H. Rostron had consulted with Bruce Ismay, who had climbed into Collapsible C and survived under circumstances that were to haunt him all his life.

It was agreed that *Carpathia* would return to New York. Halifax was closer and considered but then rejected in favour of the original destination.

Unlike today, there was no rescue coordination centre. There were no radio frequencies reserved exclusively for emergency traffic and there were no secure channels. Anyone could listen in to anyone else's wireless messages and anyone was free to interpret, or misinterpret, a message, or to garble it in retransmitting it onward.

The German tanker *Deutschland* was short of fuel and had developed control problems south of the Grand Banks early on Sunday, April 14. The Leyland steamer *Asian* had taken it in tow and was proceeding to Halifax when it heard *Titanic's* cry for help. It would appear that an *Asian* wireless message about its tow had been badly misread by someone desperate for news of survivors, and the Boston paper had run with it.

By the evening of Monday, April 15 the world knew that the North Atlantic had claimed *Titanic* and that 1,500 of its victims lay entombed in the vessel or were floating grotesquely in their lifejackets amongst the icebergs and pack ice. The tanker *Deutschland* arrived in Halifax under tow of *Asian* on Thursday, April 18 — about the same time as the survivors arrived in New York on board *Carpathia*.

To Get Results Advertise Your Wants In the Daily Globe

The Boston Daily Globe.

Advertise Your April 19th Sales In The Daily Globe

VOL LXXXI—NO 106. BOSTON, MONDAY EVENING, APRIL 15, 1912—SIXTEEN PAGES. PRICE TWO CENTS.

EVENING EDITION—7:30 O'CLOCK

TITANIC GOING TO HALIFAX.

FIND AUTO IN MARSH CASE

Rifle in Car Sought Since Lynn Murder.

Cartridges, Too, in Machine Found in Boston.

Owner Gave Name of Willis Dow—Police Seek William A. Dorr.

PHOTOGRAPH OF THE GIANT LINER TITANIC TAKEN IN ENGLAND LAST WEEK.

CAPTAIN OF TITANIC.

CAPT E. J. SMITH.

Damaged Liner's Passengers Taken Off—Come by Rail.

N Y, N H & H Notified to Provide Sleeping Cars For 600 and Coaches for 800.

BOSTONIANS ON TITANIC.

WIRELESS DISAGREEMENT.

NARROW ESCAPE AT START.

OLYMPIC WIRED SOONEST.

INSURED FOR $5,000,000.

AID HALIFAX CAN SEND.

RAIN

READ THE ADVERTISEMENTS IN TODAY'S GLOBE.

CASTLE SQUARE HOTEL

HOTEL HOLLIS

WILL LAND AT HALIFAX

N Y, N H & H R R Officials Notified To Prepare Five or More Trains.

PASSENGERS ALL SAFE

Indications That Everyone Will be Easily Rescued From Titanic

These Figures Are Facts:

They Tell the True Story

Much has recently been published to indicate that the BOSTON & MAINE RAILROAD is doing little to IMPROVE ITS PHYSICAL CONDITION. It is now in order to state A FEW FACTS.

BOSTON & MAINE RAILROAD

By the time this April 15 evening edition was released, the trains that were headed for Halifax to pick up passengers had been recalled.

Chapter 3

After The Disaster

Mackay-Bennett

In 1912 about a dozen marine telegraph cables connected Canada and Newfoundland to Europe. The French cable company, La Compagnie Française des Câbles Télégraphiques (CFCT), had operated out of Halifax with its cable ship *Contre-Amiral Caubet* since 1906, but late in 1911 the vessel had developed serious engine problems. It was lying at its berth at the Liverpool Wharf in March 1912 when the PQ cable from Brest, France, to the island of Saint-Pierre off the coast of Newfoundland, broke. About 60 nautical miles east of Cape Spear, Newfoundland, out on the Grand Banks, the cable had been "ridden by iceberg" and had gone out of service on March 21.

Model of the recovery ship Mackay-Bennett *as she appeared a few days after arriving on site, built for the Maritime Museum of the Atlantic by Thomas M. Power. Bodies on the foredeck are draped in canvas; those on the aft deck have been placed in coffins. A burial at sea is being prepared on the starboard side.*

CFCT had to act to recover its revenues, and for the third time it was forced to contract with the Commercial Cable Company to provide the services of cable ship *Mackay-Bennett* from its downtown Halifax wharf. *Mackay-Bennett* left on March 30 and repaired the break directly, but Captain Frederick H. Larnder (not "Lardner" as it is often spelled) of *Mackay-Bennett* and Captain Louis M. LeMarteleur, the CFCT client representative on board, noted on the cable repair sheet that the iceberg "is still a menace to this cable."

When the *Titanic* disaster struck, and the false hopes of the vessel and its passengers coming to Halifax were dashed, speculation in the Halifax newspapers turned to which local vessels could go out in search of bodies. As they had an agent in Halifax, it was logical that the White Star Line would make use of the variety of vessels available in this port, only about 700 nautical miles (1300 km) from the *Titanic* loss site. The cable ship *Mackay-Bennett*, the steamers *Seal* and *Florizel*, along with the government coastal vessel *Lady Laurier*, were all apparently considered. The minutes of the weekly operations meeting of the CFCT in Paris record that they released *Mackay-Bennett* from its CFCT obligations because of "*la grande émotion causée par le désastre du* Titanic."

Mackay-Bennett was contracted by the White Star Line on the evening of April 16 at a rate of $550 US per day. The following morning was spent in a frantic rush to get

Opposite: Canvas-wrapped bodies on board Mackay-Bennett*'s forecastle.*

Gloves of Canadian Titanic *victim Charles M. Hays.*

ready. A hundred tons of large blocks of ice were put in the forward hold and 125 coffins organized by Snow & Co. Ltd., the region's largest undertaking firm, were stacked on the wharf. John R. Snow Jr. joined for the trip, along with an undertaker-stonecutter named George P. Snow who normally worked next door to the funeral home on Argyle Street, in the George A. Sanford & Sons Nova Scotia Steam Marble & Granite Polishing Works. Canon Kenneth Cameron Hind of the Anglican All Saints Cathedral also went along as a clergyman to administer to the dead and to bury them, as required, at sea. The local newspapers noted that all these activities on the wharf were closely attended by "a corps of New York, Boston and local press representatives with kodaks slung on their shoulders and ready at a moment's notice for human interest pictures." The photographers would not be so welcome, however, on the vessel's return in just under two weeks time.

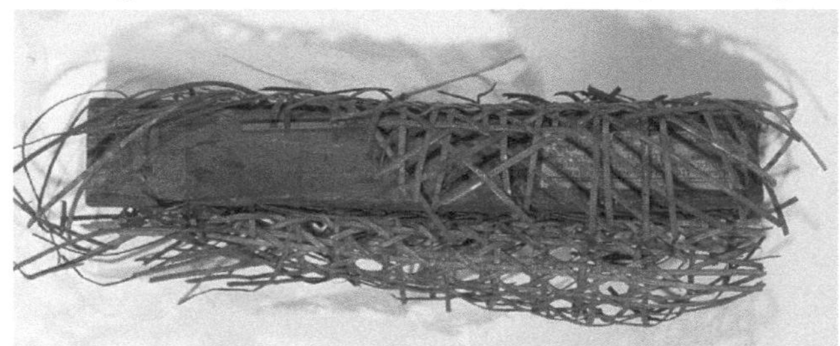

Fragment of caning from a Titanic *chair used as a pattern to repair the deckchair at the Maritime Museum of the Atlantic.*

The cable ship was off to sea shortly after noon on Wednesday, April 17. There were sporadic reports back to shore by wireless and it very quickly became apparent that there were far more bodies floating in the ocean than anyone had expected. Captain Larnder was to later describe the scene, "Like nothing so much as a flock of sea gulls resting upon the water ... All we could see at first would be the top of the life preservers. They were all floating face upwards, apparently standing in the water." Few had drowned or died of other injuries. The majority of the victims had died quickly and relatively painlessly of hypothermia from the 28°F (-2°C) water. It did not take long for White Star Line officials onshore to conclude that a second vessel would be required.

The Western Union (formerly the Anglo American Telegraph Company) cable ship *Minia* was contracted to depart very late on Monday, April 22. It too had been out on a cable repair in the vicinity of Cape Race, Newfound-land, when *Titanic*'s distress call had electrified the ether in the early hours of April 15. It had begun to make ready to go to the scene when it was evident that other vessels were much closer. *Minia* arrived back in Halifax for a rapid turnaround to go right back to sea as the second ship to assist in the collection of *Titanic*'s victims.

Meanwhile, on *Mackay-Bennett* the body count mounted. Each time wreckage and bodies were encountered, at least two of the ship's cutters with a five-person crew were put over the side. Each boat would bring back up to nine bodies in calm waters, four or five otherwise. Fifty-one bodies were recovered on Sunday, April 21, their first day at the scene. Twenty-six were recovered on Monday. A watch was maintained from the bridge and the forecastle head; flags and whistle signals were used to direct the cutters to bodies. On Tuesday, April 23, 128 bodies were brought on board. Bodies were stacked everywhere.

Only 125 coffins had been brought and embalming supplies for just 70 persons were available. The decision was made to bury many bodies at sea; those that could not be identified from their personal effects, such as pocket contents, letters, tickets, initials on their clothing, and so on, and occasionally those who could not be embalmed

Undertaker John R. Snow Jr. and ship's officer pose on the Mackay-Bennett *wharf on April 17, 1912 as coffins are loaded.*

because of injuries received in the sinking were committed to the deep in evening ceremonies conducted by Captain Larnder and Canon Hind. On the Sunday, 24 were buried at sea, on Monday, 15 more. None were buried on Tuesday, April 23 as supplies of canvas were low. A rendezvous with the liner *Sardinian* of the Allan Line Company was set up by wireless and at 7:00 p.m. Captain Robert McKillop transferred all the canvas that he had on board to assist *Mackay-Bennett* in covering or wrapping the bodies of victims. On Wednesday 77 more were buried.

Frederic A. Hamilton, Chief Electrician, recorded the scene as the ship lay silent "wallowing in the great rollers":

> *The tolling of the bell summoned all hands to the forecastle where thirty bodies are to be committed to the deep, each carefully weighted and carefully sewn up in canvas. It is a weird scene, this gathering ... for nearly an hour the words "For as [much] as it hath pleased ...we therefore commit his body to the deep"*
> *are repeated, and at each interval comes, splash! as the weighted body plunges into the sea, there to sink to a depth of about two miles. Splash...Splash...Splash.*

In total 116 were buried at sea. *The Casket, The Oldest Independent Journal in the Trade*, editorialised from Rochester, New York, in the June 1 issue that, "The custom of sea burials has been brought into great prominence by the recent Titanic disaster." Publisher Simeon Wile spoke out "against

A cutting board from Titanic *found floating by a* Mackay-Bennett *crew member and kept by George P. Snow, assistant undertaker.*

the barbarous use of the ocean's depths as a cemetery, noting that the National Funeral Directors' Association had lobbied the Chair of the U.S. Senate investigating committee "to include in his findings a recommendation for the abolishment of sea burials." The Senate made no such recommendation.

Each body was numbered sequentially as it came up to the rail in the ship's boat and was lifted carefully down on to the deck. The number was entered in the vessel's logs and stencilled onto a piece of canvas that was tied to each body. This identity mark was to then follow the person through all the records, onto the photograph of the unidentified corpses taken at the mortuary onshore, through the burials onshore, into the lists made up by the White Star Line, through to the disposition of personal effects and valuables, and finally carved into the Halifax gravestones over the next year. Any personal effects were put in small canvas bags stencilled with the same number.

Overturned Collapsible B lifeboat examined by a Mackay-Bennett *cutter, April 22, 1912.*

Wednesday, April 24 was extremely foggy with very poor visibility and the sea began to kick up. There were no new bodies collected that day. Instead, the crew continued their documentation of the 205 collected to date, with the last 77 burials at 12:45 p.m. Arminias Wiseman, of Shoal Harbour, Newfoundland, was a 22-year-old fireman on board who recorded his recollections at about age 69, in 1958. It was probably on this day that the following incident occurred:

> *In addition to the gruesome task of bringing on board and handling the victims, there were, believe it or not, some amusing incidents. One such occurrence happened while I was on duty, the 4 to 8 watch, when the second engineer sent one of the oilers with a message to the watch officer on the bridge. As he came up from the engine room and stepped out in the passageway, a body slid by him and almost knocked him down. He retreated to the engine room in a hurry with a report saying that a body had become alive and was going up and down the passageway. He was pretty scared. The engineer asked me to go up and investigate. I did and discovered a body had become dislodged from a pile that was in the passageway and due to the pitching of the ship, had slid down in front of the engine room door. The deck crew was notified and the body was secured.*

On Thursday, 87 more bodies were brought on board. Word had come by wireless that *Minia* was to take over the grim search. They drifted within sight of each other in the early hours of Friday until light came, when *Minia* transferred embalming fluid across to *Mackay-Bennett* and the two vessels continued the search together. *Mackay-Bennett* retrieved 14 more bodies by noon and then headed for Halifax "freighted with her cargo of woe." The ship had recovered 306 — a hundred more than anyone had expected. Only 18 were women, and there were no ship's officers. One of the first bodies recovered was that of a two-year-old child.

Arminias recalled that on this trip there was extra fortification for the task at hand:

> *The usual custom on the ship when she was out at sea working on a cable job was at about 6 o'clock in the evening the Chief Steward would come forward with a pail of English rum. Each man would receive 3 or 4 ounces in his drinking mug. On that trip each man was allowed a double ration. It was a case of putting spirits down to keep their spirits up!*

Crew of the cable ship Mackay-Bennett, *from a circa 1910 postcard found at the Maritime Museum of the Atlantic.*

On their return to shore early on Tuesday, April 30 the men of the "Death Ship" got an extra reward for their labours. President Clarence H. Mackay of the Commercial Cable Co. had wired Captain Larnder the previous evening:

> *In appreciation of the efficient services of yourself and your officers and crew of over one hundred men, in a work humanitar ian and yet entirely outside of your regular line of duty the Commercial Cable Company allows you and your officers and men double pay for the time engaged.*

The vessel tied up at Coaling Jetty 4 in the security of H.M. Dockyard in Halifax. Once all 190 bodies had been landed, Captain Larnder invited the press on board. Twenty newspaper men gathered in the ship's lounge at 11:15 a.m. and hung on every word as he and Canon Hind took them through the 13-day voyage. One newspaper compared the landing of the bodies to "the return of the *Chesapeake* after its mortal combat with the *Shannon*, the loss of *H.M.S. Tribune*, the wreck of the *Hungarian*, the *Atlantic* catastrophe, the *Thingvalla* and *Geiser* collision, ... the *La Bourgoyne* horror ... but it is doubtful if any of these ever appealed more strikingly to Haligonians than the foundering of the *Titanic*." Little did this writer know that in five-and-a-half years' time Halifax and Dartmouth would have to draw on their *Titanic* experience and use the same numbering system for bodies and the same numbered canvas bags for personal effects. This time, however, it would be for their own citizens as the 1917 Explosion killed over 2,000 people in an instant as the explosives ship *Mont Blanc* blew up. The atomic-sized blast is featured in the exhibit *Halifax Wrecked* at the Maritime Museum of the Atlantic.

Within two days of its return, *Mackay-Bennett* was back out to sea for the French cable company. The PQ cable had again broken, in virtually the same spot on April 26, perhaps due to the same iceberg as before.

Anglo-American Telegraph Co.
Cable Ship "Minia" April 27th/12.
2.20. a.m.

My darling Mother
I expect you will be surprised to receive this written on this paper but I am on watch now in the wireless room so thought it a good opportunity to write you. This is the most remarkable trip the old Minia has ever been on as we are looking for bodies from the Titanic wreck. You know I wrote that we were up North on a cable repair when we heard she had sunk. we arrived in Halifax about three days after and it was reported that we had "some of the recued on board" but we had not, and the reporters that came to meet us were disappointed. The same

alone were worth $450,000. I am s
to say that we have to go out again
bout 2 days up North the same pla
e we were in when we heard abo
tanic.   Etc. etc.
Your loving son
Francis Dyke.

MINIA

Thomas G. Dutton's 1885 painting of Minia *going to the rescue of a crippled sailing vessel. This Maritime Museum of the Atlantic image depicts the vessel before the hull was painted white.*

Under the command of Captain William George Squares deCarteret, the cable ship *Minia* continued the grim task at sea. However, the rate of recoveries was dropping off as the bodies and debris were fanning out to the east and north as they progressively came under the influence of the Gulf Stream. Rev. Henry Ward Cunningham of St. George's Anglican Church (the round church in Halifax) went along to bury the dead, if this was necessary. William H. Snow was the undertaker and was assisted by the ship's surgeon, Dr. Byard William (Will) Mosher.

On Friday, April 26, 11 bodies were found and recovered. Will wrote to his sister, Agnes, on *Minia*'s return:

> *... we have been chartered by the White Star to hunt for bodies, body snatching we call it. ... This place [Halifax] has been nothing but an undertaker's establishment for this last month ... left here Monday night, April 22nd. Took 150 coffins, 20 tons ice, 15 tons scrap iron.*

Minia's recoveries dropped to one each day for the next four days, ending with two bodies recovered on Wednesday, May 1. Will explained to Agnes that on Saturday, April 27,

> *it was blowing a gale and foggy, and it was several days before we could get a [navigation] sight ... Bodies were all scattered, never came on two together. We picked up some 150 [nautical] miles east and north from the scene of the wreck ... Our star corpse was [Charles M.] Hays, the President of Grand Trunk Ry.*

Opposite: A letter written by Minia*'s second electrician Francis Dyke to his mother, describing the retrieval of* Titanic*'s victims.*

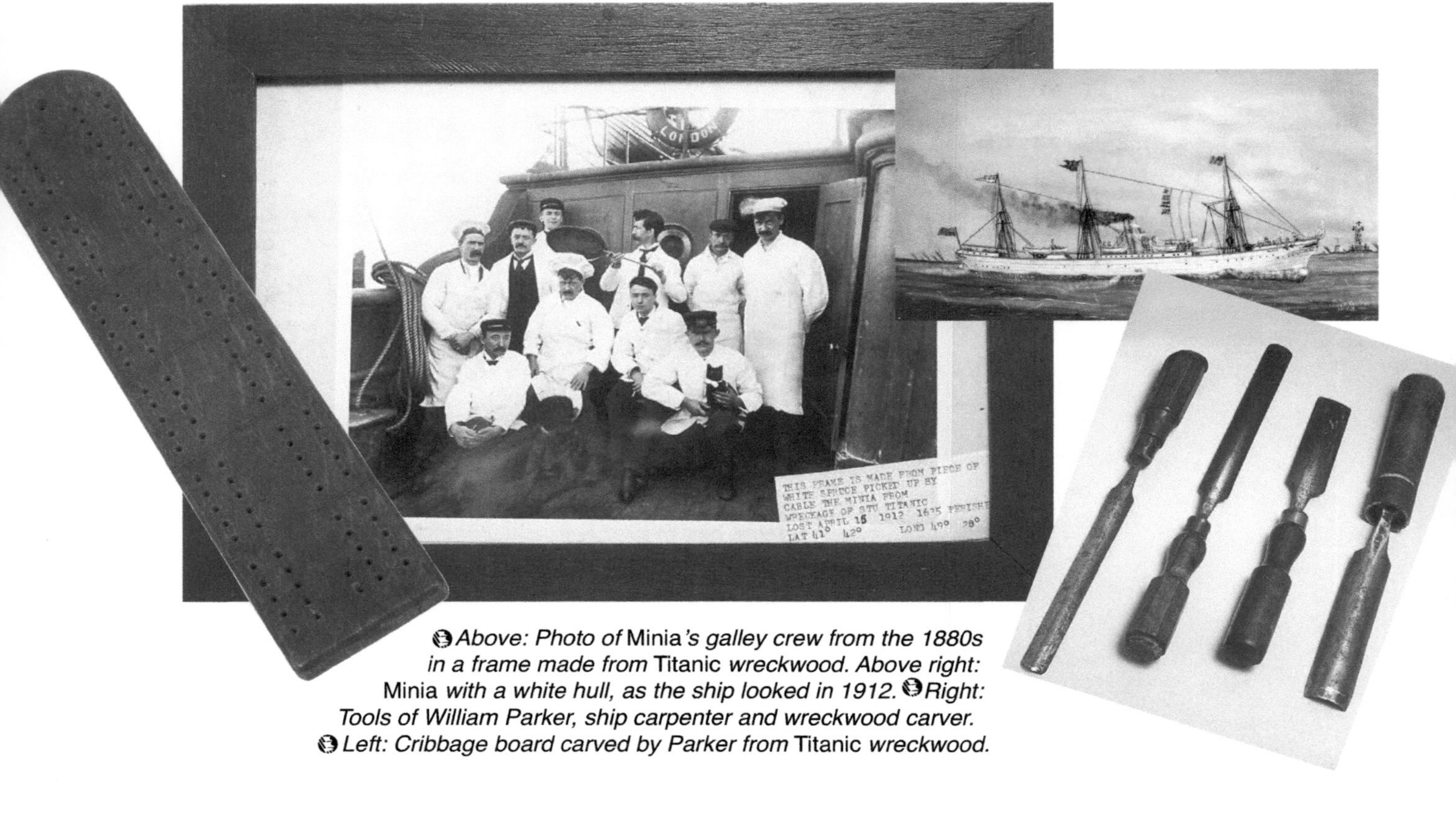

Above: Photo of Minia*'s galley crew from the 1880s in a frame made from* Titanic *wreckwood. Above right:* Minia *with a white hull, as the ship looked in 1912. Right: Tools of William Parker, ship carpenter and wreckwood carver. Left: Cribbage board carved by Parker from* Titanic *wreckwood.*

Second electrician Francis Dyke wrote to his mother on several of his early morning watches:

> *I can tell you none of us like this job at all but it is better to recover them and bury them properly than let them float about for weeks. ... I expected to see the poor creatures very disfigured but they all look as calm as if they were asleep. ... the bodies are much scattered, ... they go very fast when in the Gulf Stream — very likely many will be washed up on the Irish Coast as they are all going East.*

A boat from Minia *picking up a* Titanic *body.*

Everyone on *Minia* commented on the debris. With few bodies to find, *Minia* picked up numerous bits of wood from *Titanic*. Will wrote to Agnes,

> *Picked up any amount of wreckage. Deck chairs, doors, chests of drawers, cushions, two of the steps of the grand stairway, some beautiful carved panels (oak) and cords of painted wood, moulding, boards, etc. I am having a picture frame made from some wreckage & am looking for a picture of the ship to put in it. Also am making a cribbage board and a paper weight, Making the paper weight out of the leg of a table (oak) and am having a silver plate put on top with Titanic engraved on it. I found a card signed by Ismay in the pocket of one of the stewards whose body we picked up & am keeping it as my chief souvenir. This ship is full of souvenirs at present, everyone is making checker-boards, cribbage boards or paper weights.*

Surgeon Will Mosher standing beside a Titanic *victim being prepared for burial on board* Minia.

Francis Dyke made an astute observation in his letter home in fact noting that *Titanic* had broken up:

> *The Titanic must have been blown up when she sank, as we have picked up pieces of the grand staircase ... most of the wreckage is from below deck, it must have been an awful explosion, too, as some of the main deck planking 4 ft thick was all split and broken off short.*

Reverend Cunningham, in his report to the newspapers, noted, "Among the articles picked up were a deck chair, in good condition, a drawer from the first cabin pantry ..." This chair was donated to the Maritime Museum of the Atlantic by his family where it remains as one of the compelling artefacts from the *Titanic* disaster.

Like *Mackay-Bennett* before it, *Minia* saw what they presumed was the iceberg that "*Titanic* was supposed to have struck" and Will sent Agnes a snapshot — "We picked up one body floating alongside the berg."

Minia found only 17 bodies, buried two at sea, and finally turned for home on May 3. It arrived at the quarantine inspection station at the mouth of Halifax Harbour at 2:00 a.m. on Monday, May 6 and waited for daybreak before proceeding to the Flagship Pier in the Dockyard at Coaling Jetty 4, now rebuilt and totally internal to Jetty NJ just north of the Angus L. Macdonald Bridge. As before, no photographers were allowed in the dockyard. Will Mosher wrote, "No photos were supposed to be taken but our navigating officer managed to get a snap from the bridge and I enclose a copy."

All *Minia*'s unused coffins and embalming fluid were transferred as soon as they arrived over to the third ship that was to engage in the search, *Montmagny*.

C.G.S.

MONTMAGNY AND *ALGERINE*

Canadian government vessel Montmagny *which made two trips to the* Titanic *area to recover bodies.*

By the time *Minia* returned it was clear that there were very few bodies left floating to recover. Despite this, two other vessels were to be sent out. *Montmagny*, a Canadian government vessel, was first on May 6, followed by *Algerine* owned by Bowring Brothers Ltd. of St. John's, Newfoundland. *Algerine* left port on May 16.

Montmagny had two captains, two clergymen, and two undertakers on board, and made two trips to the area; May 6–13 and May 14–23. François-Xavier Pouliot was the regular captain, and he was supplemented by an experienced east coast skipper, Peter Crerar Johnson, who held his Masters foreign-going ticket. Rev. Samuel Henry Prince from St. Paul's Anglican Church and Father Patrick McQuillan from St. Mary's Catholic Basilica participated, as did John R. Snow Jr., the undertaker who had been on the *Mackay-Bennett*, and another from a Dartmouth firm, Cecil E. Zink. *Montmagny* was to find only four bodies, one on Thursday, May 9 and three on Friday, May 10; the first was buried at sea and the others were delivered to Louisbourg on Monday, May 13 and shipped by train to Halifax.

Opposite: Captain Peter Johnson from Halifax who presided over Montmagny *as senior captain.*

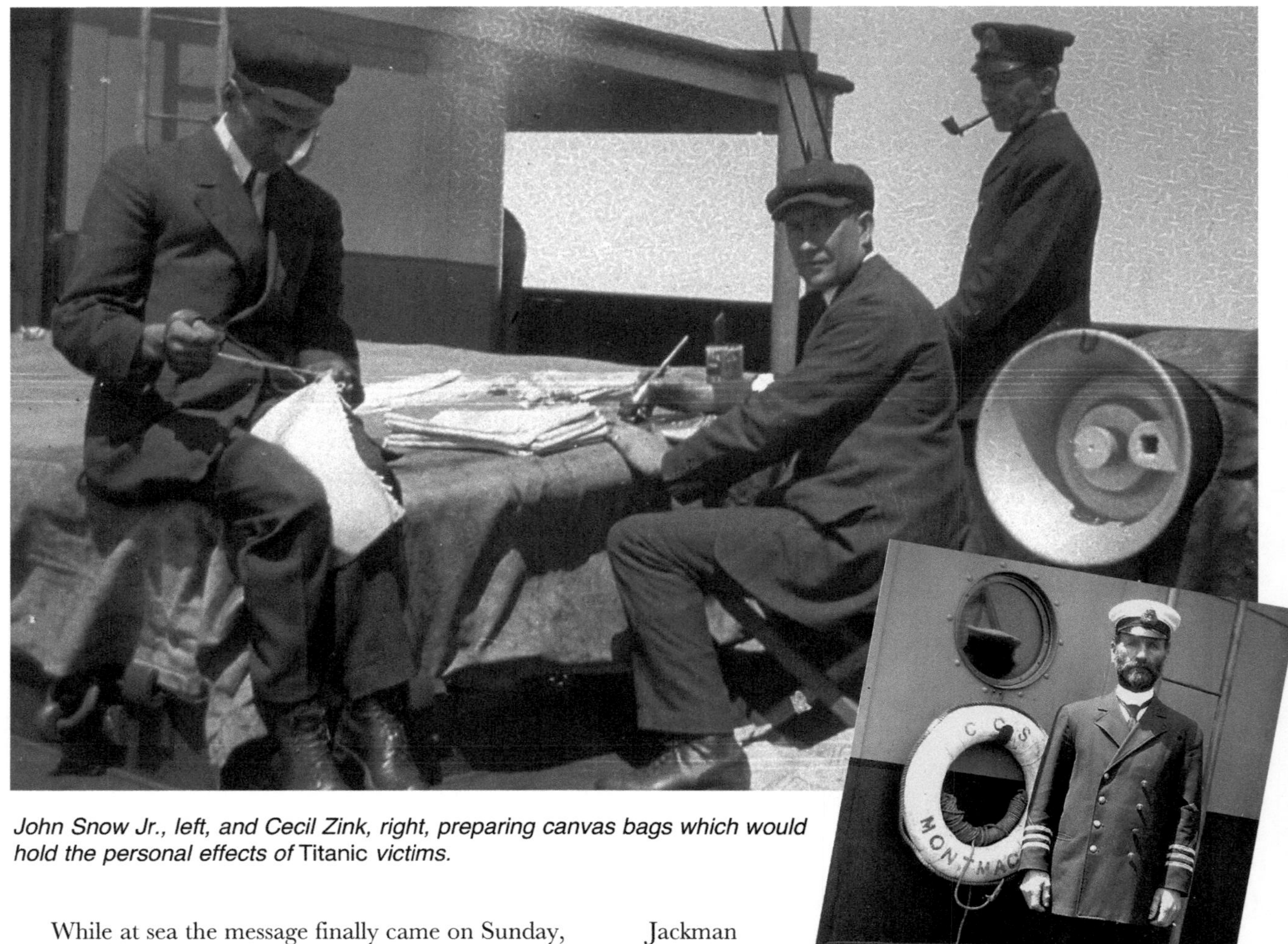

John Snow Jr., left, and Cecil Zink, right, preparing canvas bags which would hold the personal effects of Titanic *victims.*

Inset: Montmagny'*s French Canadian captain, François-Xavier Pouliot.*

While at sea the message finally came on Sunday, May 19 that all had been waiting for, "Abandon Search and Return to Halifax." It had been sent from Halifax via Sable Island and relayed by *Californian*, of the Anchor Line, which should not be confused with *Californian* of the Leyland Line that was stopped in ice on the night of April 14–15, with its wireless off, unaware of *Titanic*'s plight, just 19½ nautical miles away.

Montmagny established wireless contact with the sealing vessel *Algerine* as it came into sight and Captain Johnson conveyed details of his 13-day search to Captain John Jackman and Chief Officer Captain Giles. *Montmagny* then changed course for home, leaving *Algerine* alone with its single undertaker, Mr. Lawrence, to continue the search.

Algerine was to persist for three weeks, but only one further body was found. James McGrady, Body No. 330, a saloon steward, was recovered on about May 22. *Algerine* returned McGrady to St. John's on June 6 and he was tran-shipped to Halifax by the Bowring vessel *Florizel* on June 8,

Reverend Samuel Henry Prince saying prayers over the body of a Titanic *victim.*

arriving three days later. McGrady ended his long voyage and was interred on June 12 — the last *Titanic* victim buried in the three Halifax cemeteries.

There are no known photographs of *Algerine's* voyage. There was a camera on board *Mackay-Bennett* with at least five images surviving. Similarly there was a camera on *Minia* and perhaps as many as nine photo-graphs survive. Rev. Prince had a camera on *Montmagny* and his more than 20 photographs give a remarkable record of shipboard life, including all aspects of the vessel's *Titanic* duties. Prince kept a diary at sea and he went on to write a narrative that was published in the local Halifax newspapers on the ship's return.

Newfoundland vessel Algerine *owned by Bowring Bros.*

He also 'published' *The Montmagny Moon* on May 20, 1912, which came out "once in a blue moon," on the return trip to Halifax. It was a two-page, handwritten, light-hearted 'newspaper' that had advertisements for haircuts by the boatswain, first-class meals by the cook, embalming fluid for sale by John Snow & Co., undertakers "d'Atlantique," and P.J. Colton, the wireless operator, offered "Security, Reliability, Secrecy [sic]" for "messages taken for all parts." Under the headline "Giant Iceberg Seen," it was reported that the "white peril of North Atlantic observed at safe distance. Thought to be that which sank the *Titanic*, several photographs were taken." The one-off paper reported on Captain Johnson's oceanographic observations taken during the two voyages, detailed a near set of fisticuffs in the forecastle, and related the next probable job for the vessel. Prince was to later make lantern slides of his photographs and to present talks on the *Titanic* disaster.

Chapter 4

City in Mourning

Halifax Prepares

On Wednesday afternoon, April 17, *Mackay-Bennett* departed for her grim task. Haligonians remained transfixed by the continuing *Titanic* story as the city's five newspapers vied to carry the very latest news. On the following Sunday, churches across the nation held memorial services. At Saint Paul's Church, Halifax, a large congregation gathered, including, the Lieutenant-Governor, Supreme Court judges, and the Mayor along with representatives from such civic organizations as the Nova Scotia Historical Society, the North British Society, the Charitable Irish Society, the St. George's Society and the YMCA. The pulpit, lectern and holy table were draped in black. Before the service began, the Reverend Samuel Prince gave from the chancel steps a prayer appropriately befitting to the sad occasion:

We are met solemn in memorial today. Over the whole of the civilized world there rests the shadow of a great sorrow. Dare we do aught but drape our churches and mourn the memories of those whose bodies now lie in that place of death beneath the seas, but whose souls have passed into the presence-chamber of the king of kings. Some have been spared. Please God there may be more. We rejoice with them that do rejoice and we weep with them that weep.

The memorial service continued with the age-old Office for the Burial of the Dead.

On Tuesday, April 23, P.V.G. Mitchell, White Star's Canadian passenger agent, arrived in Halifax from Montreal to make arrangements for receiving the bodies. White Star contracted with Snow & Co. Ltd. to take charge of all recovered bodies and ordered 500 coffins and caskets from Ontario. Snow's recruited 40 embalmers from Nova Scotia and neighbouring Maritime communities. Among the New Brunswick embalmers was Annie F. O'Neil and her sister Mrs. Elizabeth Walsh of Saint John, who would embalm the women and the single child recovered.

The City and the White Star Line were faced a tremendous task, similar only to the wreck of the White Star Line's

Button from Charles M. Hays' gloves.
Postcard written from Halifax at the time of Mackay-Bennett's return with the bodies of Titanic's victims.
Opposite: Illicit photo snapped by the navigating officer from Minia's bridge.

Atlantic off Terence Bay, Nova Scotia in 1873. The *Titanic* disaster required a coordinated effort of White Star representatives, city, provincial and federal government officials, the undertakers, and the Intercolonial Railway. Arrangements had to be made from the moment *Mackay-Bennett* docked for transporting the bodies to a temporary morgue. The Mayflower Curling Rink on Agricola Street was chosen as the best building because of its size, its proximity to the landing point, the cold temperature and glassed-in viewing area. There had to be proper facilities for embalming and identifying each body prior to issuing death certificates and burial permits. Additional arrangements included the disposal of personal effects and the removal and despatch of bodies by railway, for burial elsewhere, and, for those being buried in Halifax, religious services with appropriate interment.

Mackay-Bennett had relayed by wireless to White Star's New York office the number of bodies recovered and the identifications made. By April 24, Halifax newspapers reported that 77 bodies had already been recovered and of these, 35 had been identified. Some names, however, could not be related to passenger and crew lists, while apparent misspellings and possible mistaken initials generated much distress and confusion. Relatives of *Titanic* victims, including some of the survivors who had reached New York aboard *Carpathia*, began arriving in Halifax to identify and claim the dead in anticipation of *Mackay-Bennett*'s return.

Memorial Service

FOR

Victims of S. S. "Titanic,"

Lost off Newfoundland Banks, on the night of April 14th, 1912, with more than a thousand five hundred Passengers.

At St. Paul's Church,

HALIFAX, N. S.,

Sunday, April 21st, 1912.

Order of Service used by Reverend Prince, Sunday April 21, 1912.

City papers became filled with reports on the arrival of persons such as Vincent Astor, who came in three special railway cars, in the hopes of finding the body of his father, Colonel John Jacob Astor, whose estate exceeded $100 million. The Grand Trunk Railway sent a special train for the body of its president, Charles Melville Hays. Maurice Rothschild came to recover the bodies of his friends, Isidor Straus, owner of New York's Macy's Department Store, which *was* recovered, and that of Isidor's wife, Ida, which was not found. And so the arrivals multiplied, until there were at least one hundred bereaved friends and relatives in the city.

On Friday, April 26, nearly all the "mourners" (as the press described them) then in the city gathered at the Halifax Hotel on Hollis Street for a meeting presided over by Mayor Joseph A. Chisholm. They believed that the planned arrangements for *Mackay-Bennett*'s arrival did not allow for proper privacy, nor had they been able to obtain accurate information. They formed a committee to make known their concerns. As nearly all the mourners were Americans, the local United States consul general, James W. Ragsdale, became chairman, with the mayor, the agent for the Intercolonial Railway, and C. Weston Frazee, the local manager of the Royal Bank, as members.

Mr. Mitchell of the White Star Line assured the meeting that a complete list of the dead who had been found by *Mackay-Bennett* with descriptions would be provided to the press. He indicated that the company was arranging for large burial plots at Mount Olivet and Fairview Lawn cemeteries. Particular note would be made of each unidentified body, and the coffin containing each such body, and the grave into which it was placed. Every possible mark or feature would be recorded for possible later identification. Burial by religion would be done where it could be determined. At the Halifax Hotel, the White Star Line established a bureau where mourners could obtain the latest information. Although not all the mourners' concerns could be addressed, a mutually beneficial relationship had devel-

Interior and exterior of St. Paul's Anglican Church, site of the April 21, 1912 Titanic *Memorial Service.*

oped among all parties that would prove its worth in the trying days ahead.

At another meeting of mourners, on the following evening (April 29), they discussed how identifications could be carried out with the least pain and confusion. A.N. Jones of A.G. Jones and Company, White Star's Halifax agent, informed the meeting that "certain gentlemen" had tried to bribe undertakers to hasten the embalming of friends. This attempt, however, had failed and White Star "would not stand for it." The New York press, which was involved in a bitter circulation war and had been incredibly scurrilous in its *Titanic* reporting, had "criticized adversely" the White Star Line's arrangements in Halifax for facilities to claim the bodies. On a motion by Maurice Rothschild, it was agreed, without a dissenting vote: "[W]e wish to disagree with the said adverse criticism and to express our hearty thanks for the most courteous treatment that we have received from the authorities of the White Star Line."

An extensive discussion then took place over procedures for identification. White Star stated that in some cases disfigurement would make identification probably impossible. The meeting finally agreed that when the embalmers had finished with a body, it would be brought out to the main room of the Mayflower Rink, then the name of the deceased, if already known, would be called out. Those wishing to claim the body could then come forward to confirm identification. A White Star representative assured the meeting that this was the planned procedure. The firm also issued a circular describing the process for caring for the dead, which it distributed to all hotels. Relatives and agents were required to register with A.G. Jones and Company, at 180 Hollis Street, one block north, at the Halifax Hotel.

Earlier on the Monday, at a meeting attended by

Premier E.H. Murray, Orlando Daniels, the attorney general, and the American consul general, J.W. Ragsdale, agreement had been reached for the identification of bodies and the disposal of personal effects. The government agreed to have an official at the Mayflower Curling Rink who would take charge. It appointed the lawyer R.T. MacIlreith, KC, an expert in taking evidence.

The Nova Scotia law relating to personal effects of shipwrecked victims required them to be handed over to the deceased's estate, which, with the numbers involved, could have meant lengthy delays. Because of the circumstances, MacIlreith was given the authority to transfer personal effects on satisfactory conditions being met. The effects of those buried at sea were to be sent directly to the White Star Line office in New York. In the case of American citizens, MacIlreith could hand over personal effects, after determining the legal claimants, to the American consul general who would hold them until estate proceedings in the United States had been completed and then forward them to the claimants. Over 30 cases, however, remained for months in the Provincial Secretary's department awaiting valid claim. No aspect of the government's role in dealing with the victims generated more correspondence than disposing of these remaining personal effects. Great efforts were made to obtain proper and sufficient evidence of a claimant's entitlement before dispatching any items. Claimants could not, however, be found for a few bags and these were eventually sent to the White Star's New York office.

St. Mary's Catholic Basilica, Halifax.

The government was concerned that Nova Scotian law be followed in all cases for the proper disposition of the bodies, while wishing to do everything possible to avoid administrative delays. The agreed solution was for J.H. Barnstead, deputy registrar of deaths, and the provincial coroner and medical examiner, Dr. W.D. Finn, and their staffs to establish an office on the second floor of the Mayflower Curling Rink. By having the provincial medical examiner present at the temporary morgue, it would not be necessary to hold inquests, with full formality, before issuing death certificates. This would also allow Barnstead to make out burial permits so that claimants could remove bodies without delay and arrange for their transportation. The Canadian Express Company offered to transport any bodies from *Titanic* free of charge. All the unidentified and those identified, but not claimed, would, however, have to be buried in Halifax. Religious leaders formed the Evangelical Alliance to handle the burial services.

Meanwhile, on Sunday, April 28 at churches throughout the city clergy read the White Star Line's request that when *Mackay-Bennett* arrived, people would stay away from the dockyard area. Newspapers reported on memorial services held in 12 city churches. At All Saints Cathedral, the congregation sang a special hymn, composed by the English novelist, Hall Caine, entitled "The Titanic," and sung to the tune "O God Our Help in Ages Past." All day Monday, Halifax waited expectantly for *Mackay-Bennett*'s return. For the mourners, many anxious hopes rested on its efforts.

Monument maker's 'signature' on body 188's stone in Mount Olivet Cemetary.

CARGO OF DEATH

Tuesday, April 30 dawned with a quiet grey sky. Then in the early morning the signal man on Citadel Hill announced sighting *Mackay-Bennett*. Tolling church bells told of her sighting. Flags all over the city and in Dartmouth were at half-mast. Stores had draped their windows in black and placed in them pictures of *Titanic*. Shortly before 9 o'clock, *Mackay-Bennett* could be seen slowly steaming up the harbour with "her afterdeck piled high with coffins and on her forward deck a hundred unshrouded bodies," as reported by the *Halifax Herald*. As the vessel neared George's Island the sun emerged from the clouds. Not a breeze appeared and the water remained as smooth as glass as she glided towards the Dockyard. Crowds massed on Citadel Hill, along the waterfront and gathered on roofs of hotels, public buildings and any other vantage point. Hundreds more crowded in the vicinity of the Dockyard.

Sentries at the closely guarded gates allowed entry to officials and registered mourners. A stream of mourners arrived in closed carriages, automobiles and taxis. Snow & Co. Ltd. and other undertakers with their hearses passed into the Dockyard. For the *Acadian Recorder*'s reporter the sight of the "sable clad equipages was an impressive sorrowful one, and gripped the hearts of all present, bringing before the imagination the scene of the terrible disaster, with all its grief and anguished details."

H.W. Higginson on board Minia *lifting canvas to reveal victims.*

At 9:30, *Mackay-Bennett* moored at the Flagship Pier. All present stood silent, with heads bare. The Press Association had a telegraph apparatus at the Dockyard entrance from where its reporters sent wireless messages to the outside world as events unfolded. The afternoon newspapers would print the following numbers:

Total bodies recovered 306
Number brought to Halifax 190
Number identified and buried at sea 56
Number unidentified and buried at sea 60
Number identified and brought to Halifax 65
Number not identified and brought to Halifax 125
Number embalmed at sea 106

The unloading of the 190 bodies took three-and-half hours. The last bodies were those of the first-class passengers, all of which had been identified, embalmed and placed in coffins. Reporters from 15 Canadian and US newspapers were admitted on board for Captain Larnder's 11:15 a.m. news briefing where he answered questions. Canon Hind commended *Mackay-Bennett*'s crew, noting how "earnestly and reverently all the work was done, and how nobly the crew acquitted themselves," in a task that was "hard and straining on the mind and body." John Snow Jr. was quoted in the *Evening Mail*, "There was awful evidence of the fierce struggle for life, hands clutching wildly at clothing, faces distorted with terror. But it is no use to try to describe what we saw. To do so is impossible. As I said, ours was a sickening task."

However his comments were severely countered by the ship's surgeon, Dr. Thomas Armstrong, in a statement published verbatim in the *Morning Chronicle* on Thursday, May 2. Dr. Armstrong noted, "with the exception of about ten bodies that had received serious injuries their looks were calm and peaceful, in fact, so peaceful that it was difficult to realise that they were dead." Snow denied the comments attributed to him in the same article.

Stacked coffins on the stern of Mackay-Bennett.

Meanwhile, 30 hearses delivered the bodies to the Mayflower Curling Rink, though 10, including that of John Jacob Astor, apparently went directly to Snow's. At the temporary morgue, the arrangements had transformed the interior so that it was almost unrecognizable as a curling rink. Organizers had constructed a wooden partition at the western end, behind which there were 34 embalming benches. Once embalmed, bodies were brought out into the main rink floor, to one of the 67 enclosed cubicles, which were each large enough for three coffins. The observation room or deck had been turned into a waiting room for mourners. One of the dressing rooms had become a temporary hospital, staffed by nurse Nellie Remby. Adjoining the waiting room, staff had furnished a nearby room with writing materials.

On the rink's upper floor, government officials had established an office. From there they worked to ensure legal formalities were adhered to and authentic identifications made before any bodies were released for burial. Staff made out identification cards for all bodies, giving sex, estimated age, hair colour, marks such as tattoos, and description of clothing. No viewing could take place until the undertakers had finished embalming all the bodies, placing the embalmers under great stress. When Frank Newell, an embalmer from Yarmouth, unexpectedly found the body of a relative, Arthur W. Newell, a Boston banker and a first-class passenger, he collapsed from the shock. Once viewing began, great care was taken to ensure only *bona fide* claimants could view bodies and personal effects. By Wednesday morning 11 bodies had been released to relatives. By the end of Wednesday, May 1, relatives had shipped out of Halifax 26 bodies for burial elsewhere. Ultimately 59 of the recovered *Titanic* victims were sent away for burial.

BURIAL GROUNDS

Forty-one of the bodies sent out of Halifax went to the United States, conforming to the American custom of bringing the dead home for burial; 7 bodies went to other Canadian cities, including 5 to Montreal, 8 to England, and 1 each to Norway, Italy and Uruguay.

J. Bruce Ismay had his private secretary, W.H. Harrison, and his favoured deck stewart, Ernest E.S. Freeman, whose bodies *Mackay-Bennett* had recovered, buried in Halifax. After a brief evening service on Wednesday, May 1, at All Saints Cathedral, attended by seven men, most of whom were the White Star Line's representatives, Harrison's body was taken away immediately for burial at Fairview Lawn Cemetery. His was the first Halifax burial of the *Titanic* dead. Ernest Freeman was not buried until May 10.

Because of the advanced state of decomposition of some bodies, it was decided not to attempt embalming, but to place them directly in caskets. White Star contacted the Evangelical Alliance to arrange for the early burial of these bodies to take place on Friday, May 3. On the evening before, the authorities made preparations for the burial of these mostly unidentified bodies. Four were determined to be Roman Catholics and therefore to be interred at Mount Olivet Cemetery. Rabbi M. Walter came to the morgue and recognized nine bodies as being Jewish. These nine, the authorities separated for interment in the Baron De Hirsch Cemetery. From the original 59, this left 46 for burial the next day at Fairview Lawn Cemetery.

Mass burial of 36 of Titanic*'s victims at Fairview Lawn Cemetery, May 3, 1912.*

The Evangelical Alliance chose Brunswick Street Methodist Church as the site of the Friday morning memorial service for "the unidentified many," attended by the lieutenant-governor, officers and petty officers of HMCS *Niobe*, senior army officers, and White Star representatives. Sailors from *Niobe* acted as ushers. As a solemn mark of grief, the church was draped in black and purple. In front of the pulpit had been placed the Union Jack and the Stars and Stripes. Near the pulpit were clusters of pink and white carnations, a gift of Mrs. Hugh R. Rood of Seattle, Washington. Mrs. Rood lost her husband on *Titanic* and had come from Seattle in hopes of recovering his body. She was not to be rewarded for her trip but stayed for the first services — and perhaps was waiting for *Minia*'s return on May 6, in vain as it turned out — and she contributed flowers to various services, in memory of "the dear ones who died so bravely."

In a moving sermon, Principal Clarence MacKinnon of the Presbyterian College at Pine Hill spoke in tribute of the "unidentified" who would not be forgotten and how

Photo of Titanic *graves at Fairview Lawn Cemetery,* circa *1913. Inset: Burial cards for four bodies interred at Mount Olivet Catholic Cemetery, Halifax.*

"in Halifax these graves will be kept forever green. All we know is that they heard the order: 'All men step back from the boats,' they stepped back and they shall sleep in a hero's grave."

As at All Saints Cathedral on the previous Sunday, the congregation sang, with the accompaniment of the Royal Canadian Regiment Band, Hall Caine's special *Titanic* hymn. At the service's conclusion the band played the "Dead March" from Handel's oratorio *Saul.*

Principal MacKinnon added, almost prophetically, "They shall rest quietly in our midst under the murmuring pines and hemlocks, but their story shall be told to our children and to our children's children."

While this memorial service was in progress, the bodies of those to be buried were transported from the Mayflower Rink to Fairview Lawn Cemetery for interment at 3 p.m. When the large crowd assembled at this hour, 10 of the coffins were missing but the service proceeded, regardless. At the site a long trench had been dug and all the dead, in each of the 36 numbered coffins, laid side-by-side in a common grave. One hundred men from *Niobe* lined the grave site. A large number of people — newspapers reported as many as 800 — attended the burial service conducted by clergy of various denominations. At the service's close, the Royal Canadian Regiment Band accompanied those assembled in singing of "Nearer, my God, to Thee." After the service, the grave was filled in and over each coffin's location was placed an upright wooden slab bearing the identification number. This procedure ensured that, if it became necessary, the remains could be exhumed. One Roman Catholic was later identified and removed and reburied at Mount Olivet Cemetery on May 15.

At Saint Mary's Basilica, a solemn requiem mass for the dead was held at 9.30 a.m. for the four unidentified Roman Catholics, all females. The congregation filled the large building and "in its churchly shadow were to be noted many tear filled eyes." His Grace Archbishop Edward McCarthy was on the throne. The Reverend Rector Dr. Foley preached the sermon in which he paid-tribute to the "heroic dead" in whose memory the service was being held. After the service H.L. Monaghan and J.L. Powers, members of the White Star Line's Boston office called at St. Mary's Glebe to express their great satisfaction at the requiem mass and the rector's eloquent and touching words. The interment took place in the afternoon at Mount Olivet Cemetery, attended by White Star representatives. The graves were similarly marked as at Fairview Lawn Cemetery.

That Friday also witnessed the closing of Nova Scotia's legislature, usually accompanied with much military pomp and ceremony. On this occasion the legislature rose on a respectful, sombre note. This was believed to be the first time in Canadian military history that the troops carried no arms or colours. Nor were any military bands present.

Installation of stones at Badon De Hirsch Cemetery, November 1912 Frank Fitzgerald (l). Rabbi Jacob Finegold (c), Fred Bishop (r).

Meanwhile, the 10 missing coffins had been removed to the Jewish cemetery, just up the hill from the Windsor Street gate of the Fairview Lawn Cemetery. Inquiries found that Rabbi Rev. M. Walter had earlier opened the coffins awaiting burial. In the belief that these 10 bodies were Jewish, he and leading members of the Jewish community had taken the coffins to the Baron De Hirsch Cemetery. His intention was to inter the bodies for which he had burial permits along with the additional 10 before sundown, the next day being the Sabbath when they could neither dig the graves nor bury the dead.

As the coffins taken from Fairview that Friday all had burial certificates for that cemetery, the authorities ordered them returned. At Baron De Hirsch Cemetery, the Friday interment went ahead for nine bodies considered Jewish, including that of Frederick Wormald. The next day, Saturday, provincial and White Star officials had the 10 coffins returned to the Mayflower Rink pending further inquiry. By the following Tuesday, they had been able to determine that all were Christians. The only burial on Saturday, May 4, was that of "an unknown child," one of the first bodies recovered by the *Mackay-Bennett* crew.

On Friday, May 3, it had proved impossible to bury all the unidentified bodies lying at the Mayflower Rink. On the following Monday, a burial service at Fairview Lawn was held for another 33 of the dead. With the addition of the 15 bodies sent from *Minia* to the rink on May 6, a total of 205 had now been returned to Halifax, from the 323 victims recovered. At Mount Olivet Cemetery, services were held on May 8 and May 10 for 13 of the dead. On May 10, as well, the last large burial took place when a further 32 bodies were interred at Fairview Lawn.

As no one came forward to claim two of three bodies brought to Louisbourg by *Montmagny* on May 13—a Syrian girl and another identified as J. Smith—permits were issued for their burial. On the afternoon of May 20, the Syrian girl was buried at Mount Olivet, and J. Smith at Fairview by Dean Crawford of All Saints Cathedral. The third body, identified as Harold Reynolds, was retained at the rink in the hopes that contact could be made with relatives. He was buried at Fairview on May 22.

On June 6, *Algerine* returned to St. John's with one body on board, that of James McGrady, which was shipped to Halifax for burial. All Saints Cathedral Registry of Services for June 11 records that at Evensong, the service was given for James McGrady, "the last body recovered from *Titanic* wreck." He was buried at Fairview the next day, the last of the *Titanic* burials in Halifax.

Titanic *uniform button held by the Maritime Museum of the Atlantic, having been recovered by one of* Minia's *crew from a* Titanic *victim.*

FOREVER GREEN

James McGrady proved to be the 150th Halifax burial of *Titanic*'s dead. Fairview Lawn Cemetery became the final resting place for 121, Mount Olivet Cemetery for 19 and the Baron De Hirsch Cemetery for 10 of the 209 bodies recovered and brought to Halifax. Fifty-nine bodies had been claimed and sent off for burial elsewhere. Another 119 bodies had been buried at sea by the recovery vessels specifically sent out by the White Star Line and a further 9 were found by other ships (*Carpathia*–5, *Oceanic*–3, and *Ilford*–1) and buried at sea.

After coming to Halifax to assume responsibility on behalf of the White Star Line, P.V.G. Mitchell purchased burial plots. In the case of Fairview Lawn Cemetery, White Star paid $846.75 for 3,600 square feet. Mitchell also contracted with the well known land surveyor, F.W. Christie, to plan a layout for burial plots and memorials. The Fairview area lay on sloping ground, towards the west side of the cemetery. For the mass burial of the unidentified on May 3, Christie ran a linear trench line. Thirty-seven were buried within this row, including at the north end the separate graves for William Harrison, and the "unknown child." It became the first of four rows with the successive rows slightly curving to conform with the slope's contours. Five of the dead were buried in the second row, including that of James McGrady, indicating that this was likely the last row opened. Thirty-one were buried in the third row, and 48 in the fourth row. In the case of Mount Olivet, where the *Titanic* plot was level, the layout was not stepped and the graves were placed in two rows. At Baron De Hirsch, the ten graves were terraced and laid out in two rows to conform with a planned improvement for this cemetery.

Gravestone of James Dawson, trimmer on board the Titanic, *Fairview Lawn Cemetery.*

White Star awarded Frederick Edward Bishop of Halifax Marble Works the contract for the gravestones and other stone and concrete work about the plots. After the graves had been filled in, Bishop put in the concrete coping along the heads of the graves as a base for later placement of the marker stones. The coping was put down three to four feet below ground level, with about 6 inches showing above and 14 inches across. Whether this was F.W. Christie's intention or not, for an individual standing at the bottom of the grave site, the rows of stones give a striking sense of ship's bow with its starboard side sheared away.

By mid-October 1912, Bishop had begun to receive and to letter a small quantity of the polished 'black granite' stones with bevelled tops, the quarry being unable to provide the whole lot at once. The headstones were installed in the Baron De Hirsch Cemetery first and are believed to

Titanic *graves at Fairview Lawn Cemetery. Inset: Mount Olivet Cemetery, with* Titanic *gravestones in background.*

have been in place by November 6, 1912. On the bevelled top of each stone was inscribed the deceased's name, if known, the date of *Titanic*'s sinking, April 15, 1912, and the corresponding body number. Francis (Frank) Lewis Fitzgerald, grandfather of Walter Fitzgerald, first mayor of the Halifax Regional Municipality, elected in 1996, did the lettering. Ismay paid for special stones for his secretary William Harrison and for Ernest Freeman. On Freeman's stone No. 239, he had inscribed: "He remained at his post of duty, seeking to save others, regardless of his own life and went down with the ship."

The White Star Line paid for all the expenses relating to the burials at all three cemeteries and the upkeep until 1930, when it created a $7,500 fund with the Royal Trust Company of Canada for perpetual care. Initially there were difficulties because Royal Trust would only pay for grass cutting. Until further funds were made available, Fairview Cemetery Company paid the cost of planting tulips in the fall and other flowers in the spring. Although Halifax City assumed responsibility in 1944 for the Fairview Cemetery, not until the 1980s did negotiations begin for the transfer of the trust fund and its proportional division among the three cemeteries. These negotiations finally concluded in the mid-1990s and in 1998 improvements were made to all three cemeteries that included resetting the gravestones in new foundations and adding signs and a pathway.

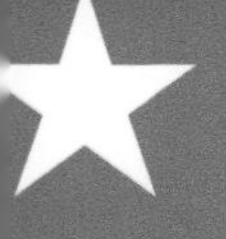

Chapter 5

PASSENGERS
THE PUZZLE OF IDENTITIES

The men of *Mackay-Bennett* really did not know what they were in for. Yes, they had a large number of coffins on board, tons of ice, undertakers and a clergyman, but they were not prepared for the scene that met them on the night of April 20-21. There was an air of considerable misapprehension as the two cutters were lowered at first light on Sunday, April 21, to begin the search.

The first body found was that of a fair-haired boy aged about 10–12 years with Danish coins in his pocket, one of the two van Billard children who had boarded third class in England and who were lost along with their father. The third body was that of a 28-year-old woman with "JH" on her chemise, not identified until 1987 as Jenny Lovisa Henriksson, from Sweden.

Body No. 4 was "a child of 2 or 3 years, a boy with fair hair and no lifejacket." He was dressed in a "grey coat with fur on collar and cuffs; brown serge frock; petticoat; flannel garment; pink woolen singlet, brown shoes and stockings." His mother had had time to dress him warmly for the trip to the upper decks. The coroner's files continue, "no marks whatever, probably third class."

Above: Gösta Leonard Pålsson, at about age two.
Opposite: One of the Maritime Museum of the Atlantic's most popular and evocative exhibits is this pair of child's leather shoes, believed to belong to Body No. 4, the "unknown child." In 2011, together with DNA evidence, the shoes helped confirm the child's identity as Sidney Leslie Goodwin.

It was surprising that the body even floated with no lifejacket for the six days since *Titanic*'s loss. This recovery, more than any other, moved the men of the cable ship. He became known as "the unknown child." It was reported in the Halifax *Morning Chronicle* on May 1, "The little body floated up alongside the searchers' boat, and it was tenderly taken on board. The sight of this little form floating face upwards on the deep brought tears to the eyes of many of the hardy sailor men." They vowed that if the child was unclaimed, they would take responsiblity for his burial.

Being able to claim a body generally meant being wealthy enough to pay for a relative to come to Halifax to make the identification and to pay for shipment of the coffin to relatives in Europe, America or Canada.

In the case of Col. J.J. Astor a private railway car was sent to fetch his remains while Charles M. Hays' corpse was taken away in a special railway car called *Canada*.

However, if you were a relative of a third-class passenger lost on *Titanic,* the chances are you may not have even

Nils, Alma and Stina Pålsson, circa *1910.*

received notification that your relative had been identified until the interment had already occurred. Certainly most relatives of third-class passengers and of crew members could not have afforded the cost of back and forth wireless messages or of a trip to come and identify a relative. The photos taken in Halifax of the unidentified bodies could not have possibly even reached the White Star offices in New York and in Southampton, England before the interment of the unidentified began on Friday, May 3.

The Wormald family, a mother and six children, of Southampton, came on *Olympic* via New York to claim their husband and father, saloon steward Frederick Wormald, in Halifax. They were refused entry to the United States, having "no visible means of support" and returned empty-handed on the same vessel. Wormald, an Anglican, was mistakenly identified as Jewish and was buried on May 3 as Body No. 144 in the Baron De Hirsch Cemetery. To add to their sorrow, on arrival back home, the family found that they had been evicted for non-payment of rent.

In the Mayflower Curling Rink, the body of the "unknown child" was laid out on a bed of roses in a small white casket which was placed symbolically in the centre of the coffins of the unclaimed women — as if they were his caregivers in death. When no one came forward to claim the child, the crew of *Mackay-Bennett* kept their vow. A number of the crew were from Halifax and were probably members of the congregation of St. George's Anglican Church, the round church on Brunswick Street. In the absence of Rev. H.W. Cunningham, who was at sea aboard *Minia*, Canon Kenneth Hind conducted the service for the child, which was attended by the 75 crew and officers of *Mackay-Bennett*. The small casket, heaped with flowers, was carried by six men to the horse-drawn hearse. Flowers had come from all over, including the White Star Line, the coroner, the undertaker John R. Snow, Jr., his wife, Mrs. Hugh R. Rood, the children of Bloomfield School, the Nova Scotia Nursery, and a bouquet signed only, "The boys from the morgue." Hundreds lined the streets as the hearse travelled from the church to Fairview Lawn Cemetery.

In May 1912 this "unknown child" was tentatively identified in a document from White Star Line as being Alma Palsson's youngest son, Gosta. It is presumed that when she finally dressed her four children and reached the upper decks, perhaps guided by Steward William Denton Cox (grave 300) in one of his several forays into third class to guide passengers to the boats, the last boat was gone. We suspect that she eased the children's fear with some music from the mouth organ. In the end she pocketed the instrument, asked another Swede, 27-year-old August Edward Wennerström, to hold the baby boy, Gösta Leonard, and did her best to comfort the other children against their rising fear. As the tilt increased, the bridge tore loose and the wave swept August into the ocean. The child was swept from his grasp and lost. August and Carl Olaf Jansson swam together and found Collapsible A, struggled into the partly water-filled boat, then later were saved by Boat 14. Gösta was lost.

Alma had been found by *Mackay-Bennett* as the first body on Thursday, April 25 and was buried on Wednesday, May 8. The records do not tell us whether her husband Nils even had a chance to claim her remains.

Various candidates for the identity of the "unknown child" have been put forward. In 2002, DNA testing indicated a new one, Eino Viljami Panula, born March 10, 1911, to Juha and Maria Panula of Finland. But further DNA testing in April 2011 confirmed the body was in fact that of nineteen-month-old Sidney Leslie Goodwin from England. The entire Goodwin family, en route to Niagara Falls, where Sidney's father had been offered a job, perished in the disaster. Today, the headstone of the "unknown child" stands as a tribute to all the children lost on Titanic — the "ship of dreams."

GEORGE WRIGHT

Had George Henry Wright's body been found by the Canadian ships that went out to recover bodies, it would have ended up in the funeral parlour of Snow & Co. Ltd. at 90 Argyle Street in Halifax. As it was, some of the remaining contents of his home at the southeast corner of Young Avenue and Inglis Street ended up at the Argyle Street premises of the well-known auctioneer Melvin S. Clarke, following the sale at the house on May 30, 1912.

Wright, a wealthy bachelor, was returning from Europe as a first-class passenger on RMS *Titanic* and little is known of his time on the vessel. He was said to be a heavy sleeper and to retire early, and it is possible that he slept through much of the disaster. Wright was known to be a philanthropist and active in various causes related to temperance and morality. The *Halifax Herald* made the announcement of his death on April 20, 1912, "Halifax loses good man in Geo. Wright" and a special mention was made the following day at the service at St. Paul's Anglican Church in the Grand Parade. One month later, aboard *Montmagny*, Rev. Samuel H. Prince of St. Paul's held a remembrance service for Wright over the loss site.

Wright was born in 1849 on the Dartmouth side of the Halifax harbour, in Wright's Cove. *The Evening Mail* of 1897 wrote that he "was Intended for a Halifax Business Man, But Went Abroad and Made a Fortune, While Still a Young Man."

At age 17 he travelled to the United States and conceived the idea of a trade directory and gazette, often referred to as "Wright's World." A shipper's guide to the United States followed, then a guide for the Australian exhibits at the 1876 Philadelphia World Exposition. Wright then set out on a four-year series of travels in

Portrait of George Wright, a wealthy Halifax businessman, by the Halifax Notman Studio.

Australia and the Far East during the compilation of his massive 4,000-page, *Australia, India, China and Japan Trade Directory and Gazetteer, embracing Canada, South and Central America, the West Indies and Africa*. The first edition, in 1880, was followed by at least four more under various titles. Wright returned to Halifax in early 1896 to settle down and to sail his yachts *Alba*, then *Princess*. The Wright Cup competition is still held at the Royal Nova Scotia Yacht Squadron in Halifax, over a hundred years after he presented it in 1898.

In Halifax, working with local architect James C. Dumaresq, Wright devoted himself to developing two fine buildings on downtown Barrington Street that still anchor its Victorian streetscape at the southwest corner

Left: A Wright house, South Park Street, Halifax. Right: George Wright residence, 989 Young Avenue, Halifax.

of Prince Street— the St. Paul Building and Wright's Marble Building. Wright and Dumaresq also developed a striking series of seven Victorian homes facing South Park and Morris Streets, with 21 other residences behind these on Wright Court and Wright Avenue.

In 1902 he built for himself the fine home found today at 989 Young Avenue.

The estate auction on May 30, 1912 at the house included the following:

Particulars
Drawing Room - Upright Cabinet Piano, Music Cabinet, Chairs, Lounges, Tables, Quill Work Tables, Bric-a-Brac Cabinet, Rare and Costly Ornaments and Bric-a-Brac, Valuable Oil Paintings and Engravings, Curtains.
Beautiful Collection of Persian Carpets and Rugs, Comprising 3 Carpets and 19 Rugs of the very choicest of the Orient, selected for the late Mr. Wright by an expert and connoisseur, the most valuable collection east of Montreal.
The Den - 1 Marlow Repeating Rifle, the best that can be obtained; 1 Shot Gun, cost $100; other Guns and Revolvers. A unique collection of Curios and Relics gathered in all parts of the world.

One of Wright's 4000-page trade directories.

Additional items were advertised in the newspapers of May 31 and June 6 when there was still the large oil painting left and five of the rugs and a horse:

A beautiful Bay Gelding, Morgan bred, formerly owned by late George Wright, 9 years old, 1150 lbs, one of the finest

Left and middle: Two more of Wright's South Park housing developments. Right: St. Paul's Building on Barrington Street. Below: Historic plaque on Wright's Young Avenue residence.

road horses in the Province, will road 12 miles an hour; will be sold without reserve.

Wright's estate of almost $250,000 was left to his family and to various causes. His home went to the Women's Council and $20,000 was left, with $10,000 for maintenance, "for a building to be erected for the purpose of bringing the people together ... to provide clean amusement in order to check the lure and bad influence of the streets." This bequest withstood two court challenges until the YMCA successfully sprung the funds in 1952 to greatly assist in building their new structure on South Park Street. In recent years, the auditorium where the brass memorial plaque to Wright was hung became a men's locker room, but in late 1996 the plaque was relocated to the lobby area.

There is a memorial stone to Wright erected by his brother in Christ Church Cemetery in Dartmouth and a plaque in the Women's Council House, though perhaps his buildings serve as the best testament to the man and to his success in life. Almost all of these have stood the test of time, withstanding the 1917 Explosion in Halifax Harbour and the more recent developments in the city's South End.

ALL WALKS OF LIFE

Marital disputes are private matters that are probably best kept private. Occasionally, however, they erupt into the public eye. So it was with the troubles of Michel and Marcelle Navratil (*née* Carretto). Michel was a Slovakian emigré who had settled in Nice, France, in 1902 and became a successful tailor, married the Italian Marcelle, and started a family. The sons Michel M. ("Lolo") and Edmond Roger ("Momon") were almost age 4 and age 2 in April 1912. The parents had separated, perhaps because Mr. Navratil resented the interference of his mother-in-law in his marriage.

The boys had been to visit their father over the Easter weekend of 1912; at the end of the weekend all three were gone — a story all too familiar today. The father had kidnapped them and, with the help of friends, had gone to nearby Monte Carlo, taken the alias Louis M. Hoffman, and sailed for England to board *Titanic* for America. At the last moment during the sinking, Louis Hoffman passed the two boys through a ring of crew protecting Collapsible D just before lowering, the last boat to be launched. They arrived on board *Carpathia* as the "Titanic Orphans." Margaret Hays of New York, a survivor, spoke French quite fluently and took the boys under her wing and into her New York home.

*The "*Titanic *orphans," in New York, April 1912. Edmond 'Momon' (left) and Michel 'Lolo' (right) Navratil.*

Body No. 15 was found on Sunday April 21, *Mackay-Bennett*'s first day of searching. It was Louis Hoffman, who was described as:

MALE. ESTIMATED AGE, 36. HAIR & MOUSTACHE BLACK. CLOTHING - Grey overcoat with green lining; brown suit. EFFECTS - Pocket book; 1 gold watch and chain; silver sov. purse containing £6; receipt from Thos. Cook & Co. for notes exchanged; ticket; pipe in case; revolver (loaded); coins; keys, etc.; bill for Charing Cross Hotel (Room 126, April, 1912).

Hoffman ended up in the Mayflower Rink with the others and lay unclaimed until May 15 when he was buried as the last body placed in the Baron De Hirsch Jewish Cemetery.

The orphans quickly became babies of the media and their photographs went around the world. Marcelle recognised her boys in a French newspaper and set out for New York on the White Star Line's *Oceanic* to reclaim them from the New York Children's Aid Society. It was in mid-ocean that she and the other passengers on board had a sobering reminder of the *Titanic* tragedy. The White Star liner *Oceanic* came upon Collapsible A of *Titanic* with three bodies. The crossing was interrupted while the bodies were buried at sea and the waterlogged lifeboat lifted on board.

Marcelle was reunited with her two sons three days later and returned with them to France on *Oceanic*'s May 18 return trip. The father apparently told the older boy, as a message for Marcelle, that it had been his intention to get settled in America, then send for her to join them. As it was, she had the job of raising the two boys as a widow in Nice. The younger boy grew up to become an architect and builder. His health suffered as a result of time in a German prisoner-of-war camp in the Second World War, and he died in 1954 at the age of 43.

The older brother, Michel, studied philosophy, married, and earned a PhD. He worked in a university in France for many years. He had not known of his father's grave in Canada until the late 1980s. In August 1996, at the age of 88, Prof. Navratil arrived in Halifax on board the cruise ship *IslandBreeze* and saw his father's grave for the first time. Though he was buried as Louis M. Hoffman on May 15, 1912, by the time the White Star Line installed the gravestones in early November 1912 his true identity was sorted out and his stone was lettered as "Michel Navratil." When his son went into Baron De Hirsch Cemetery on August 25, 1996 and rested his hand on the gravestone, he later reported to David Flemming, the director of the Maritime Museum of the Atlantic in Halifax, that he could hear angels singing and the lullaby that his father used to sing to him as a child some 84 years ago, before that fateful April night.

Another survivor, Hilda Slayter, had been pursuing a singing career in England for three years before returning on *Titanic* to British Columbia, where she was to be married. She was born on April 5, 1882, and had grown up, along with ten brothers and sisters, in a house on fashionable Argyle Street, Halifax. Her father purchased the house in 1871 from his medical partner, Dr. Alexander Hattie. In this house Dr. William B. and Clarina Slayter raised their family and William had his medical practice. Of their children, two became doctors, one in private practice and the other in the army. Two sons became lawyers, one an admiral in the Royal Navy, and another an architect. Of Hilda's sisters, two married army officers and a third a gentleman rancher.

Hilda Slayter claimed to be "one of the last off the ship." When the ship struck the iceberg, she was advised to go back to bed as they were in no danger. Half an hour later she came back on deck. After she and others had assisted each other in putting on life belts, she was "grabbed by a man who passed me along a row of men who were standing at the rail, and was carried along the deck so rapidly that I nearly lost my footing. I was put into boat 13, just as it was being lowered." One of the few boats to leave filled to capacity, with 65 occupants, Lifeboat No. 13 went shakily down but when it reached the water, the crew could not release the falls since the sea was so calm. Normally the ocean swell raises and lowers a lifeboat as it reaches the sea and the motion pulls out enough of the lowering rope to allow the crew to undo the hook holding the falls to the lifeboat, but this night it was glassy calm with no swell. Worse still, boat No. 15 was descending directly above and threatening to capsize No. 13. At the last moment, two crewmen cut the ropes and No. 13 drifted away just before No. 15 hit the water. Later, Hilda Slayter graphically described *Titanic*'s end to a Vancouver newspaper:

Photo of Hilda Slayter, circa 1911.

> *We had nobody to row except the cook's boys, and although they didn't know anything about it, worked manfully. We were clear of the ship about three-quarters of an hour when she sank. She was aglow with lights right up to the last and was plainly visible. As the lights went out and she sank under the water we heard a loud moaning wail, and she disappeared from view. Almost immediately there came a explosion, and then a second. Somebody shouted from another boat, "hold tight for a big waves", but no big waves came; the sea was as calm as a millpond.*

Hilda survived but the trousseau she brought with her from England did not. The lifeboat was taken on board *Carpathia* and Hilda was reported by the Halifax newspapers as safe in New York. For a few days interviews appeared, then a photograph. She proceeded on to British Columbia and the wedding to Henry R.D. Lacon of Denman Island went ahead in Vancouver on June 12. She lived on the west coast for over fifty years and revisited Halifax in 1964, then died the following year and is buried in the family plot in Camp Hill Cemetery in Halifax. The family home on Argyle Street went through a number of uses, including a hotel annex and a printing shop. It still stands as a heritage building and has served as the St. Paul's Anglican Church Parish Offices for more than 20 years.

Among the other *Titanic* survivors who developed connections with Nova Scotia was the wealthy Fortune family of Portage Avenue, Winnipeg. The European trip had been planned in part to keep daughter Mabel separated from a jazz musician who had not won her parents' favour. The father, Mark, and son Charles perished in the wreck. The mother, Mary McDougald Fortune, and three young daughters survived in Lifeboat 10. Mabel went on to marry the musician. Ethel also married and her son Crawford later headed A.V. Roe while it built the first swept wing AVRO Arrow aircraft. Alice married Charles Holden Allen and together they regularly visited Chester, Nova Scotia, and came to love the area. They were both buried in St. Stephen's Anglican Church graveyard in Chester, he in 1955, she in 1961.

Titanic carried a significant number of Lebanese, or Syrian, emigrés who were seeking a new life in North America; there were perhaps as many as 88. Most boarded in Cherbourg and were cited as "Syrians" on the third-class passenger lists. Among them were Nicola Yarrid, a miller from the small village of al-Hakoor, about 100 kilometres northeast of Beirut, and his two children, Jamila, age 12, and Elias, age 10; five other members of the family, including the mother, had already emigrated. Isaac, an older brother, had married Mary Shediac of Yarmouth, Nova Scotia, and was living in Liverpool, Nova Scotia, running a fruit and confectionery business on Main Street; he had already anglicized his name and was known as Isaac Garrett.

Hilda Mary Lacon's (née Slayter) gravestone in Halifax's Camp Hill Cemetery.

The Yarreds show up on the White Star passenger list under the surname Nicola — probably a ticket agent's error listening to a Lebanese father give his name —Nicholas. Fate intervened and the father could not board *Titanic* because of an eye infection, so the children went on their own with another Syrian couple watching over them; they were a party of 13, including an uncle. Only the children survived, perhaps with the help of John J. Astor who was recalled as having assisted one of the children into a lifeboat.

Isaac and his brother-in-law met them in New York and they all returned to Yarmouth on the steamer *Boston* on April 24, and thence to Liverpool where the children lived above Isaac's store. Their stay in Canada was relatively brief. Eventually Jamila and Elias became Amelia and Louis Garrett and were reunited with their parents in Jacksonville, Florida. In later years, when Louis wrote about his *Titanic* experience, the brief sojourn in Nova Scotia was never mentioned.

THE "UNSINKABLE" MOLLY BROWN

The July 17, 1920 sailing of a nondescript, United States Shipping Board wooden steamer, *Quinneseco*, from Sydney in northern Nova Scotia, with a cargo of 3000 tons of coal bound for Denmark, via Hull, England, brought a *Titanic* survivor to Halifax. Margaret Tobin Brown (Maggie to her friends and Mrs. James J. back in Denver, where her wealthy husband lived) was on her way to England on vacation with her nieces, Helen Tobin and Mrs. William Harper (of the U.S. publishing company of the same name). Maggie, or Molly Brown, almost 45 in 1912, had been among the first-class passengers who survived the *Titanic* disaster, leaving the ship early in portside Lifeboat No. 6 which was nominally in command of quartermaster Robert Hichens. He turned out to be less than resolute and Molly effectively took command, organising the other women and teaching them how to row. Molly Tobin was born on July 18, 1867, in Hannibal, Missouri, and had learned to row on the Mississippi as a child.

Molly Brown (right) and her nieces in Fairview Cemetery.

On board *Carpathia* and in New York, Molly addressed the sick and used her knowledge of French and German to work with the third-class survivors. She remained in New York to assist. On *Carpathia*'s next visit to New York, on May 29, Molly organised the presentation of a 'loving cup' to Captain Arthur Henry Rostron and medals to all members of the crew to recognise their role in the flawless rescue at sea.

Why Mrs. Brown and her nieces had chosen a bulk coal carrier rather than a more conventional liner to travel on in 1920 is a mystery. What is known is that she had another brush with a marine misadventure aboard *Quinneseco* and found herself in Halifax. The July 26 Halifax newspapers give a brief description of the events, after the vessel's arrival,

... after a narrow escape from destruction from burning bunkers. At one place the flames had eaten into the sides of the [wooden] ship to within about an inch of the water. ... 700 miles out [from Sydney, Nova Scotia] fire started in the port bunker from some unknown cause [circa July 20]. With every device at their command the crew fought heroically with the flames and in about two days had them extinguished. The ship was capable of proceeding but with heavy weather prevailing it was decided that with her side almost burnt thru and in a charred state it would be unwise to proceed. Accordingly she was headed for Halifax, arriving here safely. [night of July 24]

A week later the "unsinkable" Molly Brown was reported in the newspaper as visiting "the last resting place of the unfortunates who met their end on the ill fated liner, to commemorate the heroism and sacrifice of that world tragedy." On August 6 having spent three days making floral tributes, Molly placed a wreath on each of the 150 *Titanic* graves in Halifax. After two weeks in Halifax, waiting for repairs to *Quinneseco*, Molly Brown and her nieces were again on their way to Europe on another vessel, *Biran*.

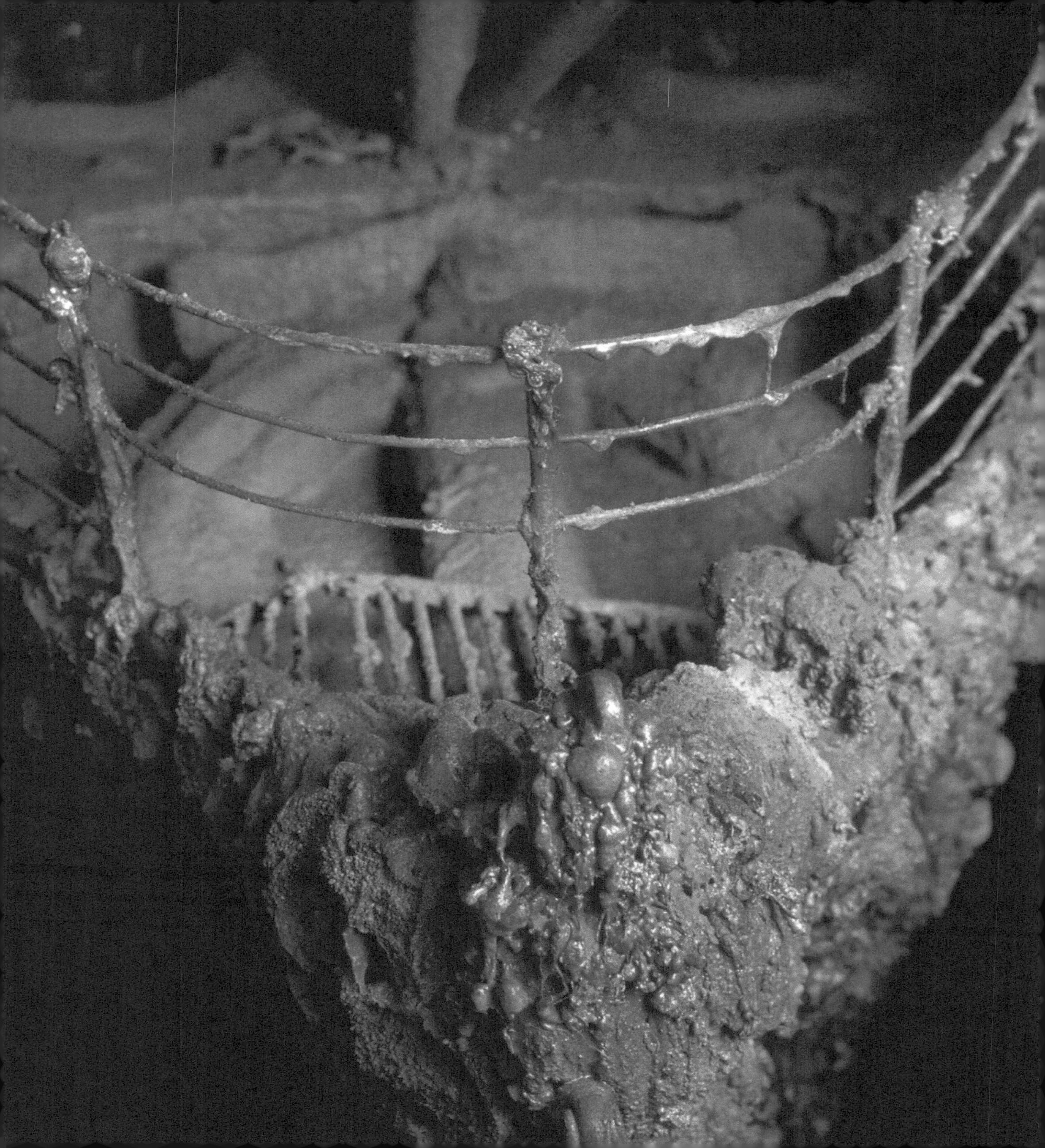

Chapter 6

THE STORY LIVES ON

FASCINATION AND DISCOVERY

After the *Titanic* disaster, American and the British authorities immediately held inquiries, gathering testimony from dozens of witnesses. This material provided the g rist for the retelling of this tragedy. Before the end of 1912 at least seven books had appeared and in 1913 an eighth was published. Their titles — *Story of the Wreck of the Titanic, The Sinking of the Titanic* (used twice), *A Titanic Hero, The Deathless Story of the Titanic, The Truth of the Titanic* and just *Titanic* — left no doubt as to their subject matter. Halifax's role was given a chapter in several of these 'instant books.' In the 1930s Newfoundlander E.J. Pratt wrote a poem on the disaster.

The next surge in popular interest came in 1953 when Hollywood first sank the liner in a film entitled *Titanic.* In 1955, Walter Lord's book, *A Night to Remember,* was on the bestseller list for six months and three years later was released as a film. This marks the start of a continuous fascination with *Titanic* that has persisted up to the present. Titanic Enthusiasts of America formed in 1963 and evolved into the Titanic Historical Society Inc., dedicated to the history and memory of RMS *Titanic* and the two sister ships. In 1983, the society's magazine, *The Titanic Commutator,* devoted a full issue to Halifax with the engaging title "Footsteps in Halifax, Footnotes to History." This was the first serious work ever compiled from the coroner's records at the Public Archives of Nova Scotia. Authors John P. Eaton and Charles A. Haas gave attention to the three cemeteries and the need to research the still unidentified dead.

Two views of Titanic *represented by Maritime Museum of the Atlantic models: as built (above) and bow of the wreck covered with rusticles (opposite).*

After the 1985 discovery of the wreck of *Titanic,* a split brought about a second rival group, Titanic International Society, Inc., including Eaton and Haas. This society's members researched the identities of twenty-two of the victims including that of Jenny Lovisa Henriksson and a young immigrant, Vendla Maria Heininen, who wore a shirt with the initials "V.H." and had 150 Finnish marks sewn into her clothing. In what can be described as the first large 'pilgrimage' to the Halifax *Titanic* sites, members of the society organized a ceremony in 1991 at the cemeteries where the inscriptions for six new identifications were unveiled.

The fascination with *Titanic* began to centre around the search for the wreck during the 1980s. American media accounts give the impression that the wreckage was discovered by one particular scientist at the Woods Hole

Oceanographic Institution (WHOI) on Cape Cod. In fact, it was a team effort and depended on earlier research, like the evidence in Peter Padfield's 1965 book, *The Titanic and the Californian,* the first serious work based on historical facts. Numerous schemes followed, including a proposal in 1974 to the Canadian Broadcasting Corporation to use a lowered scanning sonar to locate the hulk.

The first searches with a serious chance of success were financed by oil entrepreneur Jack Grimm of Texas, in 1980, 1981 and 1983, using the deep-ocean resources of Columbia University, New York, and Scripps Institution of Oceanography, California. The third of these cruises mobilized in Nova Scotia, at the Bedford Institute of Oceanography. These efforts came up short although they did eliminate a large piece of the ocean floor as not having the wreck.

Then, in 1985, a joint Franco-American effort with co-chief scientists Jean-Louis Michel and Robert D. Ballard mounted a two-month summer commitment to the search. Michel, of the French oceanographic institute IFREMER, reviewed the seafloor swath-mapping data of Columbia and Scripps so as to avoid needless repetition of work on the areas covered by Grimm. The sidescan swath-mapping of *Le Suroit* ran steadily but came up empty-handed. The French team were unaware that on their first survey line on the northeast side of their mapped area, their sidescan sonar missed the wreck of *Titanic* by less than 50 metres and for a month thereafter their survey was working farther and farther away.

Next came R/V *Knorr* of Woods Hole with the deep-towed ARGO camera package. Michel joined Ballard on the WHOI vessel and the camera package was systematically and slowly towed over the northeast extension of *Le Suroit*'s survey area. Time almost ran out for the *Knorr* team when, in the early morning hours of September 1, 1985, the now-famous images of *Titanic*'s boilers came up the cable to the TV monitors where Michel was on the graveyard shift. An excited cook was sent to get Ballard out of his bunk. The lab erupted with joy at their accomplishment.

Since then, the exact position of the wreck has become public knowledge and various deep-ocean submersibles have visited it. WHOI in *Alvin* in 1986, the French *Nautile* for RMS *Titanic*, Inc. began salvage of artefacts in the debris field in 1987. The Russian pair of *Mir* submersibles dove to the wreck for IMAX in 1991. *Nautile* went again in 1993, 1994 and 1996, for artefact recovery. Canadian James Cameron used the *Mirs* for filming in 1995 for his blockbuster film *Titanic,* released in December 1997. *Nautile* completed its 100th dive on the wreck in August 1998, recover-

An artist's rendition of the bow of Titanic *ploughed into the sea floor with submersible* Mir *hovering above.*

ing a 22-ton hull fragment that they failed to raise in 1996. *Akademik Mstislav Keldysh*, with its two *Mirs*, dove the first 15 tourists to the wreck in September 1998.

The claiming of the wreck by RMS *Titanic,* Inc. and its salving of the debris field has spawned considerable controversy. Some take the view that this is disturbing a graveyard. The Maritime Museum of the Atlantic agrees with the International Congress of Maritime Museums' position that the for-profit U.S. firm has compromised study of the wreck site and also, by precedent, that of other marine archaeological sites. The company claims that they are recovering their investment by displaying the artefacts for public enjoyment. The corporation has sought and obtained American court orders to prevent other parties from touching or photographing the wreck, or even approaching the area. These were struck down on appeal in March 1999.

Through all this, the Canadian government has taken almost no interest. Other than a modest scientific program run in conjunction with the IMAX filming in 1991, Canada has left the wreck and its protection or salvage to the Americans and the French.

RUSTICLES AND STEEL

After *Titanic*'s bow broke off, the stern floated for perhaps as long as two minutes while every thing loose rained down onto the ocean floor 3,800 metres below. Bodies that sank reached the bottom in less than an hour. The pressure of 6000 pounds per square inch ensured that they never floated to the surface and were eventually destroyed by organic processes. The two pieces of the ship lie almost one kilometre apart, with a field of debris between them. Almost all the wood has been eaten up by benthic organisms while the wreck itself is covered in "rusticles," a term coined by Woods Hole scientists to describe the rust-like icicles, that drape down from every part of the steel on the ship. An iron rusticle is not a solid mass. It is a very fragile reticulate framework of needle-like crystals of iron minerals gathered in a series of spherical aggregates. Scientific work on these structures at the Bedford Institute of Oceanography (BIO) in Dartmouth, and at a nearby university, suggests that the rusticles depend upon bacteria for their formation. They appear to grow quite quickly — perhaps up to half a centimetre per year. Below the overhanging stern poop deck where the blades of the two broken propellers can be seen, the seafloor is littered with hundreds of rusticles that have been separated from the steel above by currents, fish, crabs, falling pebbles released from melting icebergs, or by bumps from submersibles and the wash from their thrusters.

Yuri Bogdanov (left) of Moscow's Shirshov Institute and Steve Blasco (right) of the Bedford Institute of Oceanography, Dartmouth, N.S., prior to the IMAX crew's 1991 dive. Left: Titanic *rusticle obtained in 1991 and displayed at the Maritime Museum of the Atlantic.*

In other places there are "rust flows" that appear to creep across the decks of the vessel, or across the sediments of the ocean floor. Photographs taken five years apart show that they are spreading up to 10 centimetres per year. The slow dissolution of the wreck provides valuable information about the microbial processes that are corroding the steel.

In 1991 the small Canadian team working from the Russian vessel *Akademik Mstislav Keldysh*, the mother ship of the two *Mir* submersibles, brought up nine samples of metal from the debris field. The hull plate fragments exhibited conchoidal fracture, prompting scientists in the federal Department of Energy, Mines and Resources to do Charpy Tests on the metal. This test records brittleness as opposed to ductility of steel when the metal is stressed quickly and the results indicated that the *Titanic* hull plates were relatively brittle in cold water.

The suggestion is that the hull steel of the time,

Photographs of the same Titanic *bench frame in 1986 (above) and 1991 (right), showing the creep of the rust flow from the wreck (which has reached the bench in 1986) across the ocean floor well beyond the bench by 1991.*

especially in waters of -2°C, was much more likely to fracture than the steel of today. This finding has opened up a whole field of speculation by marine architects and others as to the nature of the injury to the starboard bow of the liner when it struck the iceberg.

In addition to the work on the steel, some research has been done on the geology and morphology of the wreck area. The ship lies just east of a major continental margin "canyon," or incised channel, that guides sediment runoff and periodic slumps from the edge of the Grand Banks southwestward down into the Sohm Abyssal Plain at about a 5,000-metre water depth. In a 1988 scientific paper, Elazar Uchupi and others at WHOI interpreted this incised channel to connect back up the continental slope to an already-named shelf-edge canyon — Cameron Canyon! Their map of the area then prophetically shows Cameron Canyon extending right down to the Sohm Abyssal Plain and passing just to the west of the *Titanic* area. (The Cameron honoured here is not filmmaker James, but rather, Canadian scientific researcher Harky Cameron of Acadia University, now deceased.) Canada formally named this deep-ocean channel Titanic Canyon in 1991 and has named a number of the seamounts in the area after the vessels involved in the disaster, including *Carpathia* and three of the Canadian vessels that collected the victims.

Titanic is a time marker for geologists. As it was not there on April 14, 1912, all subsequent changes in the sediment distribution are the result of the wreck interacting with the bottom currents and affecting the sediment dispersal patterns. Oceanographers have always thought of the abyssal depths of the ocean as rather barren, lifeless places. The IMAX dives offered biologists, in this case Russian scientists, an opportunity to observe the area for a considerable length of time and the amount and diversity of the biota seen was greater than expected.

HALIFAX'S *TITANIC* SITES

Since Halifax's founding in 1749, the Grand Parade has been the city's civic and commercial focal core, surrounded as it is by official buildings — Province House and City Hall — and in more recent years by rising bank towers. If *Titanic* enthusiasts could imagine themselves in 1912, standing in the middle of the Grand Parade, within their gaze, or a short walk away, would have been many of the sites important to Halifax's role in the *Titanic* story. A good number of those sites are still with us. St. Paul's Church, outside and inside, looks much like it did when on the first Sunday after the liner went down many people crowded inside for a memorial service. Many knew George Wright as one of their fellow parishioners and who had given freely to St. Paul's parish work.

Fairview Lawn Cemetery, now a popular Titanic *tourist attraction.*

Across the street at 90 Argyle was Snow & Co., the undertakers, who were under contract to the White Star Line to find and organise 40 undertakers, the coffins, supplies, hearses and the temporary morgue. Snow's survives and now comprises part of a restaurant. Next door was George A. Sanford & Sons' monument works which supplied one of the undertakers for *Mackay-Bennett.* Further along Argyle was numbers 76-80, the auction business of Melvin S. Clarke who in late May and early June auctioned George Wright's elaborate estate. Further to the south, and still standing, at 1706 Argyle is St. Paul's Parish House, that had been the childhood home of *Titanic* survivor Hilda Slayter.

On the southwest corner of Prince and Argyle, in 1912, stood the old Grand Central Hotel where the Reverend Samuel Prince lived. As the curate at St. Paul's, he had taken the memorial service on April 21 and then gone out with *Montmagny* to search for more victims. Further along Argyle Street was the Halifax Marble Works, operated by Frederick Edward Bishop who received the contract to obtain, cut and letter and then install the 150 marker stones at the three cemeteries. He was known as "Stonewall" Bishop, both because of his professional work and for his hockey goal tending.

Along Barrington Street, to the south of the Grand Parade, stand two of the most important buildings in the commercial streetscape. In 1896, George Wright commissioned the Wright Building, now 1672 Barrington. For Wright and his architect, J.C. Dumaresq, this ultra-modern structure demonstrated confidence in Halifax's progressiveness, and it still bears Wright's name carved in stone. A few years later Wright commissioned Dumaresq to design, within the same block, an even larger building. St. Paul's Building stands on the corner of Prince Street at 1684 Barrington. Such was the prestige of its address in Wright's day that it had four consulates as tenants.

The best known of George Wright's successful real estate ventures lies in Halifax's south end, at the corner of Inglis Street and Young Avenue. In 1902, Wright had Dumaresq's firm design a personal residence

Titanic *kitsch: A beer bottle from 1996*

that is recognized as one of the nation's most architecturally impressive examples of the Queen Anne Revival style. In his will, made just before he boarded *Titanic* at Southampton, George Wright left this house to "the Council of Women to be used as an Institution for carrying out their work and [to] assist in suppressing other evils such as I have been writing about and trying to put down." His estate also benefitted the YMCA at 1565 South Park Street where a plaque to him can be found.

Wright also established an integrated housing development at the southwest corner of South Park and Morris streets that has withstood the test of time. The grand houses for the wealthy face South Park, and around the corner, on Morris Street, are found the more modest, middle-class homes. In behind, on a cul-de-sac called Wright Avenue, are the working-class row houses.

The waterfront of 1912 was lined with wharves; of those that figure in the *Titanic* story only one remains today. The cable ship *Mackay-Bennett* operated from a wharf of the same name, now called Karlsen's Wharf, just north of the Purdy's Wharf towers on Upper Water Street. *Minia*, of the Western Union Company, was berthed at the Central Wharf, now lost under these same office towers. The Cable Wharf at the foot of George Street was not built for Western Union until 1914. Further north, near the Angus L. Macdonald Bridge, and presently inaccessible within the Naval Dockyard, was the Flagship Pier on Coaling Jetty 4. This is where *Mackay-Bennett* unloaded its "cargo of death" on April 30, 1912. *Minia* followed to unload six days later.

Hilda Slayter's childhood home on Argyle Street, Halifax.

The churches of 1912 still stand, although the Brunswick Street Methodist Church (now the United Church) suffered a disastrous fire in 1979. All Saints Cathedral is on Tower Road at College Street, while the Roman Catholic Saint Mary's Basilica, with its granite spires, is seen at Barrington Street and Spring Garden Road. St. George's Church, the round church at the corner of Cornwallis and Brunswick streets, was selected by the officers and crew of *Mackay-Bennett* for the memorial service for the "unknown child." In the same area, at 2660 Agricola, once stood the Mayflower Curling Rink that was used as the temporary morgue.

No visit to Halifax's *Titanic* sites can omit the three cemeteries where 150 of the recovered bodies are buried. Fairview Lawn Cemetery is best reached by following Connaught Avenue and entering at the prominent gates on Windsor Street. Adjoining Fairview is the Baron De Hirsch Cemetery where 10 victims lie buried. Mount Olivet Roman Catholic Cemetery can be reached from the other two cemeteries via Joseph Howe Drive and Dutch Village Road to Mumford Road.

All the original coroner's files and related White Star Line lists of the *Titanic* victims are housed at the Public Archives of Nova Scotia at 6016 University Avenue. These files include victim descriptions, lists of personal effects, correspondence, as well as a few related photographs and published material. This research collection is available on microfilm.

This book has drawn liberally from the artefact and research collection of the Maritime Museum of the Atlantic in Halifax which holds the world's finest collection of *Titanic*'s wooden artefacts found floating after the sinking in 1912. These items, along with photographs and documents, make up the permanent *Titanic* exhibit at the museum, "*Titanic*: The Unsinkable Ship and Halifax."

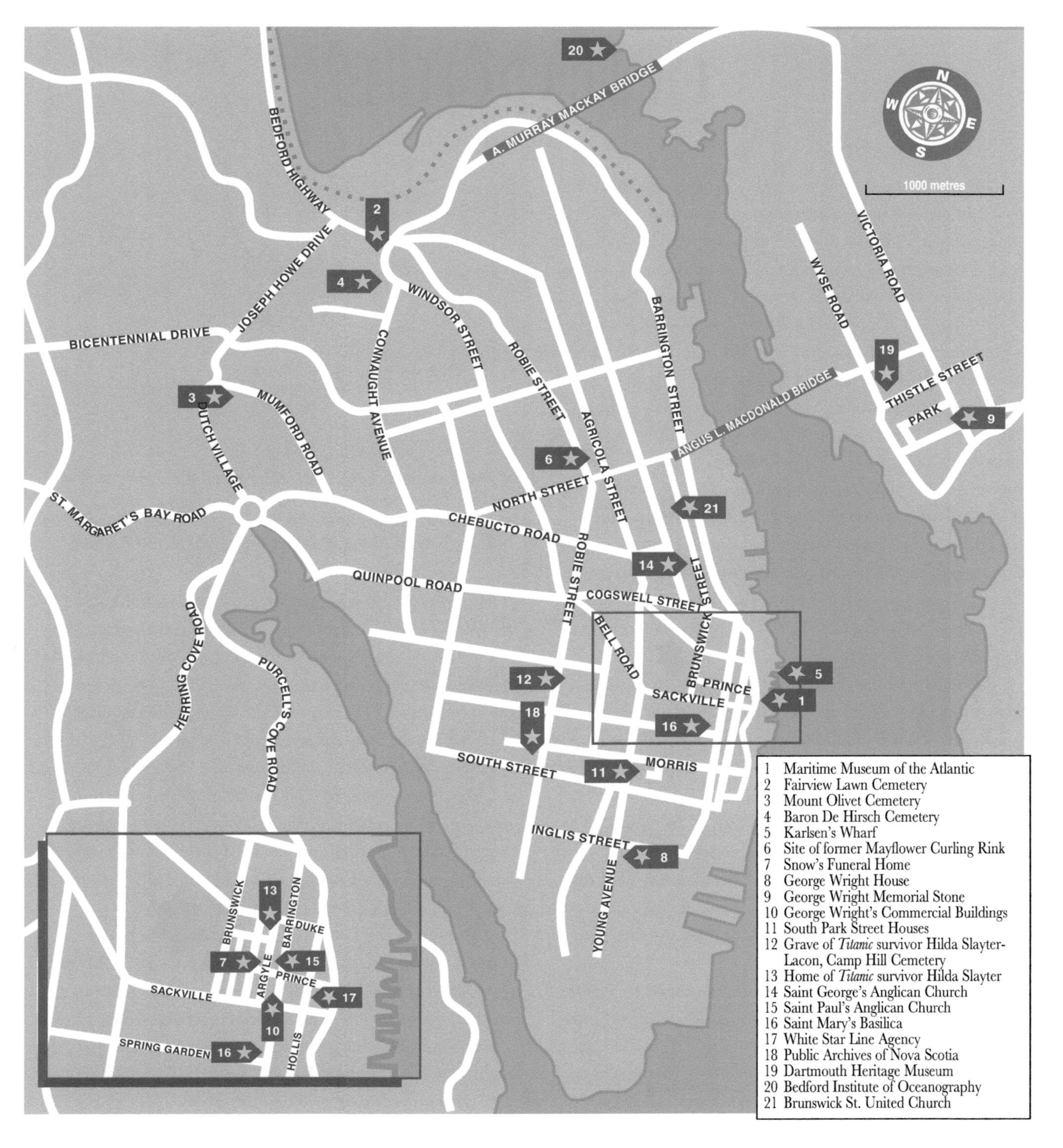
A. MURRAY MACKAY BRIDGE
BEDFORD HIGHWAY
JOSEPH HOWE DRIVE
BICENTENNIAL DRIVE
WINDSOR STREET
CONNAUGHT AVENUE
ROBIE STREET
AGRICOLA STREET
BARRINGTON STREET
WYSE ROAD
VICTORIA ROAD
ANGUS L. MACDONALD BRIDGE
THISTLE STREET
PARK
MUMFORD ROAD
DUTCHVILLAGE
NORTH STREET
ST. MARGARET'S BAY ROAD
CHEBUCTO ROAD
QUINPOOL ROAD
COGSWELL STREET
BRUNSWICK STREET
BELL ROAD
PRINCE
SACKVILLE
HERRING COVE ROAD
PURCELL'S COVE ROAD
SOUTH STREET
MORRIS
INGLIS STREET
YOUNG AVENUE
BRUNSWICK
BARRINGTON
DUKE
ARGYLE
SACKVILLE
PRINCE
SPRING GARDEN
HOLLIS
1000 metres
N
W
E
S
1 Maritime Museum of the Atlantic
2 Fairview Lawn Cemetery
3 Mount Olivet Cemetery
4 Baron De Hirsch Cemetery
5 Karlsen's Wharf
6 Site of former Mayflower Curling Rink
7 Snow's Funeral Home
8 George Wright House
9 George Wright Memorial Stone
10 George Wright's Commercial Buildings
11 South Park Street Houses
12 Grave of Titanic survivor Hilda Slayter-Lacon, Camp Hill Cemetery
13 Home of Titanic survivor Hilda Slayter
14 Saint George's Anglican Church
15 Saint Paul's Anglican Church
16 Saint Mary's Basilica
17 White Star Line Agency
18 Public Archives of Nova Scotia
19 Dartmouth Heritage Museum
20 Bedford Institute of Oceanography
21 Brunswick St. United Church

EPILOGUE

It took about one day. It was a day of denial as the White Star Line issued brave press statements, a day of slow realisation as relatives onshore hoped for word of another rescue vessel, a day of media frustration as questions could not be answered. All to no avail — *Titanic* was lost.

By the end of the week the various ocean liner firms were triumphing their safety features and proudly showing off their crews lowering and raising their lifeboats — and quietly adding lifeboats. Miraculously, room was found on *Olympic*, *Titanic*'s sister ship, for enough lifeboats for all passengers and crew. Even before the various inquiries brought down their recommendations for lifeboat changes, it had become standard practice to have lifeboat spaces for all on board. Other changes in the lifeboat regulations required adequate provisions and compasses in lifeboats, propulsion, and regular training for vessel crews. Passengers and crew were to have a pre-assigned lifeboat stations to avoid confusion.

Wireless had proved its value during *Titanic*'s loss. Henceforth it was required on all ocean-going vessels and a watch was maintained 24 hours a day. Safety and navigational signals were now given priority over commercial traffic. The wireless operator became a regular member of the crew, responsible to the captain of the vessel. SOS became the standard emergency call and eventually quiet periods in each hour were assigned to give distress calls a better chance to be heard. Rockets fired at sea ceased to be used for signalling between vessels and became exclusively a means of signalling distress.

The Atlantic shipping lanes were modified to a more southerly route during winter months. Most importantly, the International Ice Patrol was established in 1913. Every year since, except during war years, many nations have supported regular surveillance to track ice and icebergs during winter and spring. Satellites are now being used to monitor ice and to distinguish icebergs in the pack ice.

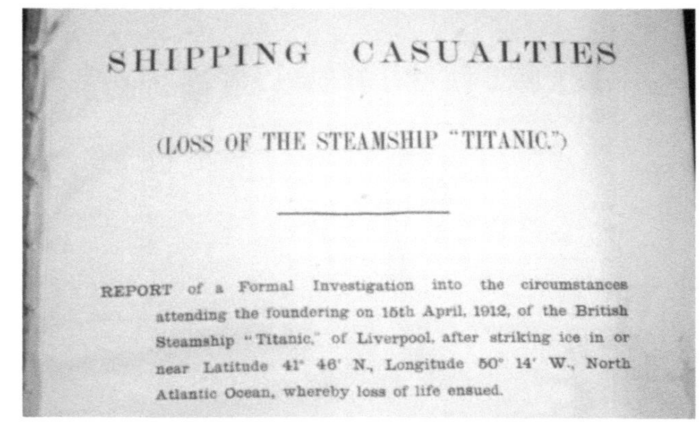

SHIPPING CASUALTIES

(LOSS OF THE STEAMSHIP "TITANIC.")

REPORT of a Formal Investigation into the circumstances attending the foundering on 15th April, 1912, of the British Steamship "Titanic," of Liverpool, after striking ice in or near Latitude 41° 46' N., Longitude 50° 14' W., North Atlantic Ocean, whereby loss of life ensued.

In the disaster's aftermath, Cunard staff referred to this leather-bound copy of the exhaustive British inquiry report to improve saftey standards aboard the Cunard fleet. It is now displayed at the Maritime Museum of the Atlantic.

Vessel design, especially the height of watertight bulkheads, came under scrutiny. Double hulls, considered by the inquiries of 1912, are still not universally installed.

One lesson that was not learned from the *Titanic* disaster was how to combat the debilitating effects of hypothermia from immersion in cold water. Not even the tremendous loss of merchant mariners and naval personnel in the North Atlantic during the Second World War brought the lesson home. That learning came only in 1982 with the loss of the semisubmersible oil rig *Ocean Ranger*.

The sinking of *Titanic* cut across all social boundaries. It did not matter how wealthy you were, or your rank on the vessel; on April 15, 1912, if you didn't get a place in a lifeboat, you were lost. The assumption that technology can always win over nature suffered a devastating blow that April night — a lesson that perhaps each generation must relearn and remember, again and again.

Sources

Expanded photograph and artefact credits can be found at the Maritime Museum of the Atlantic website: museum.gov.ns.ca/mma/titanic/titanic.htm or by writing to the author at P.O. Box 41, Halifax, Canada B3J 2L4.

Key to Abbreviations: T = Top, C = Centre, B = Bottom, L = Left, R = Right, MMA = Maritime Museum of the Atlantic, Halifax, N.S., LRT = Learning Resources and Technology, N.S. Dept. of Education, Halifax, PANS = N.S. Archives and Records Management, Public Archives of Nova Scotia Division, Halifax, UFTM = Ulster Folk and Transport Museum, Belfast, Northern Ireland, GSC-A = Geological Survey of Canada-Atlantic, Dartmouth, N.S., SHP = Rev. Samuel Henry Prince Collection of the author.

Photo Credits

Maritime Museum of the Atlantic: cover(BL), M1998.6.1A, LRT, N-24,154; p 1, poster, M1998.14.1, LRT, N-24,125; p 5, Jamie Steeves, slide collection; p 7, slide coll.; p 14(B), LRT, slide coll.; p 15(BLR), Derek Harrison, slide coll.; p 16(B), Jamie Steeves, slide coll.; p 22(T), MP400.311.1, N-19,459; p 22(B), slide coll,; p 23, M1998.6.1A & B, LRT, N-24,154 & 24,155; p 27, slide coll.; p 28(TL), slide coll.; p 28(C), Derek Harrison, slide coll.; p 29(T), MP18.109.13, N-24,132; p 31, MP29.3.2, N-9,356; p 34(TL), Derek Harrison, slide coll.; p 34(TR), M82.21.2, LRT, N-24,445; p 49(BL), Derek Harrison, slide coll.; p 53, slide coll.; p 69, Kathy Kaulbach MMA Map, revised and corrected by Formac Publishing; p 70, slide coll.; back cover(LR), Jamie Steeves, slide coll.

Alan Ruffman: p 2, SHP; p 9(BR), Courtesy of Brian Ticehurst, Southampton, England; p 10, Courtesy of Anita N. Bailey (née Edwards) and Jane E. Bailey, Halifax; p 17(L), SHP; p 18(L), 71 m-high iceberg No. 31, Atlantic Air Survey Flight No. 1, April 20, 1982; p 18(T), Courtesy of *The Mail-Star*, Halifax, July 11, 1974; p 19, Baffin 78-025; p 24, Courtesy of Ed Kamuda, Titanic Historical Society, Inc., Indian Orchard, Mass.; p 25, *The Boston Daily Globe*, Evening Edition, April 15, 1912, p 1; p 26, Harold Winard Higginson Coll.; p 30, HW Higginson Coll.; p 36, SHP; p 37, Canada Dept. of Transport, Negative No. NL 18,450 from 43rd Annual Report, Dept. of Marine and Fisheries, 1910; p 38(T), SHP; p 38(LR), SHP; p 39(T), SHP; p 39(B), National Archives of Canada, PA-77,747, John Maunder Coll., Holloway Studio, St. John's, Nfld.; p 40, SHP; p 41(CL); p 42, SHP; p 44(LR); p 46, *Harper's Weekly*, May 11, 1912, p 8(CL); p 47(T&B), single collage, SHP; p 48(T), period postcard, Novelty Mfg. & Art Co., Limited, Montreal; p 48(CR); p 57(B); p 58, A Rainbow of Time and of Space, Orphans of the Titanic, by Sidney F. Tyler, p [14]; p 61, Helen Tobin Scrapbook, Courtesy of Scott and Lisa Vollrath Coll., Helen is behind the wheel, her sister Florence Harper (née Tobin), centre; p 67(C), Courtesy of *The Daily News*, Dartmouth, N.S., September 3, 1995,p 1, Sandor Fizli.

Julian Beveridge: cover(C); p 13(CL); p 17(R); p 21(L); p 29(B); p 32; p 33; p 34(T); p 34(CR); p 41(CL); p 41(CR); p 43(T); p 43(CR); p 44(C); p 50; p 51(T); p 51(CL); p 56(T-3 photos); p 56(B); p 57(T-4 photos); p 62; p 65(CL); p 67(BR); p 68(TL); p 68(C); back cover(TL).

Other Sources: p 64, GSC-A, Edward B. MacDormand, artist, Dartmouth, N.S.; p 65(C), GSC-A, Lara Aumento; p 66(L), GSC-A, Robert D. Ballard, Woods Hole Oceanographic Institute, Woods Hole, Mass.; p 66(R), GSC-A, © IMAX Corporation/Undersea Imaging International & TMP (1991) I Limited Partnership, Toronto, Ont.; p 11, PANS N-9,189; p 34(B), PANS N-716; p 35, PANS N-715; p 55, PANS N-5,159; p 6, UFTM H1561A; p 9(T), UFTM Courtney #8; p 14(T), UFTM H1546; p 15(T), UFTM H1602; p 21(R), Claes-Göran Wetterholm Archive, Stockholm via Gerd and Hjördis Ohlsson, granddaughters of Mauritz Ådahl; p 52, C.-G. Wetterholm Archive, Stockholm; p 54, C.-G. Wetterholm Archive; p 16(T), Newfoundland Museum, St. John's, Newfoundland, Allan Clarke; p 20, Newfoundland Museum, Rose Smart; cover(R) and p. 59, late 1911 photo in England from the Alan Hustak Coll., Montreal, Courtesy of Mrs. Beatrice Lacon, West Vancouver; p 8, *Illustrated London News*, April 20, 1912, p 587, Courtesy of Ken Marschall, Redondo Beach, California; p 12, U.S. Library of Congress US Z62-26,812; p 45, Dalhousie Univ. Archives, Halifax, Thomas Raddall Coll., Photo Group FG-2-20, No. 1,150; p 49(C), Peter Douglass, Dartmouth, N.S.: Russ Lownds Coll., photo by Brewster, Nov. 6, 1912; p 27, Mrs. Bailey.

Artefact Credits

Cover(BL), MMA M1998.6.1.A; cover(C), MMA exhibition prop; p 5, MMA M81.228.1; p 7, MMA Courtesy of Mark Boudreau, Port Hawkesbury, N.S.; p 13(CL), MMA M1998.13.1.W; p 13(R), PANS RG41, Vol. 76A, No. 14, third proof, found on steward Herbert Cave, Body 218; p 14(BR), PANS AG 32.01; p 15(CR), Courtesy of Margaret F. Haley Coll., Halifax; p 15(BL), PANS AG 32.02; p 16(T), Courtesy of Peter St. Croix, St. Mary's, Nfld.; p 16(C), PANS AG 32.02; p 17(R), Maritime History Archives, Memorial Univ. of Newfoundland, St. John's, Nfld., portion of Map E.007, coll. by Prof. Keith Matthews; p 20, Newfoundland Museum, St. John's, Nfld., 997.3, pre-1935 accession; p 21(L), History Coll., Nova Scotia Museum, Halifax, Z3025a; p 21(R), Gerd and Hjördis Ohlsson Coll., Sweden; p 22(B), Parks Canada CE.36.60.34, J.S. Horton, Fort Beauséjour Coll., Aulac, N.B.; p 23, MMA M1998.6.1 A & B; p 24, Titanic Historical Society, Inc. Coll., Marine Museum of Fall River, Inc., Mass.; p 28(TL), Parks Canada CE.36.60.35, J.S. Horton, Fort Beauséjour Coll., Aulac, N.B.; p 28(C), MMA M67.61.1; p 29(B), MMA Courtesy of the children of the late Arthur and Edna Sharpe, daughter of George P. Snow; p 32, Dartmouth Heritage Museum, Dartmouth, N.S.; p 33, MMA M80.86.1; p 34(CL), PANS AG 32.04; p 34(T), MMA Courtesy of Peter Douglass; p 34(TR), MMA M82.21.2; p 34(CR), MMA M85.98.2; p 41(CL), Parks Canada CE.36.60.35, J.S. Horton, Fort Beauséjour Coll., Aulac, N.B.; p 41(CR), MMA M1998.7.1; p 42, SHP; p 48(CR), Catholic Cemeteries Commission, Halifax; p 49(BL), MMA M97.20.1; p 56(B), Dartmouth Heritage Museum, Dartmouth, N.S.; p 62, MMA exhibition prop; p 65(BL), MMA Courtesy of Steve Blasco, GSC-A; p 67(BR), N.S. Dept. of Economic Development & Tourism; p 68(TL), MMA M97.31.1; back cover(TL), MMA M1998.7.1; back cover(BR), MMA M81.228.1.

Index